WHISTL

ASTROPHYSICS AND SPACE SCIENCE LIBRARY

VOLUME 262

WHISTLER PHENOMENA

Short Impulse Propagation

Edited by

CSABA FERENCZ

ORSOLYA E. FERENCZ

DÁNIEL HAMAR

and

JÁNOS LICHTENBERGER

Space Research Group,
Eötvös University,
Budapest, Hungary

Springer-Science+Business Media, B.V.

A C.I.P. Catalogue record for this book is available from the Library of Congress.

ISBN 978-90-481-5710-5 ISBN 978-94-015-9642-8 (eBook)
DOI 10.1007/978-94-015-9642-8

Printed on acid-free paper

CONTENT

PREFACE

In memoriam dr. György Tarcsai

In space research and applications and in some modern electrotechnical, communicational applications and in other science and technology (e.g. in buried target detection) the propagation of electromagnetic signals with arbitrary shape (form) and wide band or extra-wide band spectrum has a basic role. Therefore, the importance of the real full-wave solutions of Maxwell's equations in media with complex electromagnetic signal-medium interactions has increased in the last decades. Simultaneously the measuring technics, first of all the measuring technics onboard space vehicles produced a fast development and we have a lot of signal-registratums with acceptable interpretations and a lot of signal-data without any well based explanations. The methods of signal analysis also produced a remarkable enhancement.

In space activity we have a special aspect if we see the problem of the propagation of arbitrary shaped electromagnetic signals. At the Earth and around the Earth – in the higher atmosphere etc. – and in the interplanetary space, at the Sun and at other planets of the Solar System a lot of important phenomena (atmospheric activity, lightnings, seismic-type activity, medium-fluctuations etc.) have characteristic cycles (frequencies) from the extra low frequency-bands (ELF) or very low frequency-bands (VLF) to the higher frequencies of the ultrashort waves. However, the form of these phenomena, the shapes of these signals are arbitrary, they are no strictly-periodic or monochromatic signals. Therefore, nowadays, the real full wave solutions without monochromatic, quasi-monochromatic or strictly-periodic approximations have great importance especially in the ELF-VLF bands.

In space research, the investigation of the ELF-VLF signals generated by lightnings and propagating through the Earth's plasmasphere (magnetosphere) – i.e. of the whistlers – has had great importance. This is a tool not only in the monitoring of the magnetosphere, but for example in the investigations of the seismic activity (precursors and earthquakes). In Hungary, our dear friend and colleague, the late György Tarcsai was a pioneer of the ELF-VLF research, and he was an active and successfull researcher to his early dying. We hope that this book is a memorial to George.

The methods of solution and the processes which are presented in this book are general and applicable in every frequency-bands in which the phenomenological description of the medium-signal interactions is valid.

However, originating from our space practice and from the importance of the ELF-VLF applications, the examples of applications are ELF-VLF applications in space activity, mainly applications on the propagational problems in the Earth's plasmasphere.

Budapest, 20th December 2000

Csaba Ferencz

Chapter 1

REAL FULL WAVE SOLUTION OF MAXWELL'S EQUATIONS

1.1 Theoretical background

In electromagnetic wave propagation problems the general propagation situation is independent of the signal shape. So it is identical in monochromatic and non-monochromatic cases. This general situation is presented in Fig. 1.1. In this concept the strong inhomogeneities of the medium crossed by the e.m. signal (wave) are concentrated around continuous surfaces A_i, and between these narrow surface-like regions –inside the volumes V_i– the medium is homogeneous or weakly inhomogeneous. Therefore, the fast changing of medium parameters in space – i.e. strong inhomogeneities – can be described using functionals, e.g. step-functions (Heaviside distribution). In the other parts of medium it is possible to use advanced versions of the well known solution methods of Maxwell's equations.

The general method which can manage this problem is the Method of Inhomogeneous Basic Modes (MIBM – *Ferencz*, 1978a; 1996, pp. 219-223). Using this method we can derive the propagating signals in the presence of regular inhomogeneities or distributions. If distributions are present in Maxwell's equations and the characteristics of medium are independent of time, the system of equations separates to two parts. Let us see the Maxwell's equations

$$\overline{\nabla}\times\overline{H}=\overline{J}+\frac{\partial\overline{D}}{\partial t},\qquad \overline{\nabla}\cdot\overline{B}=0,$$
$$\overline{\nabla}\times\overline{E}=\frac{\partial\overline{B}}{\partial t},\qquad \overline{\nabla}\cdot\overline{D}=\rho, \tag{1.1}$$

where $\overline{E}$ is the electric field strength, $\overline{D}$ is the electric displacement, $\overline{H}$ is the magnetic field strength, $\overline{B}$ is the magnetic induction, $\overline{J}$ is the current density and ρ is the electric charge density, $\overline{\nabla}$ is the well-known differential operator.

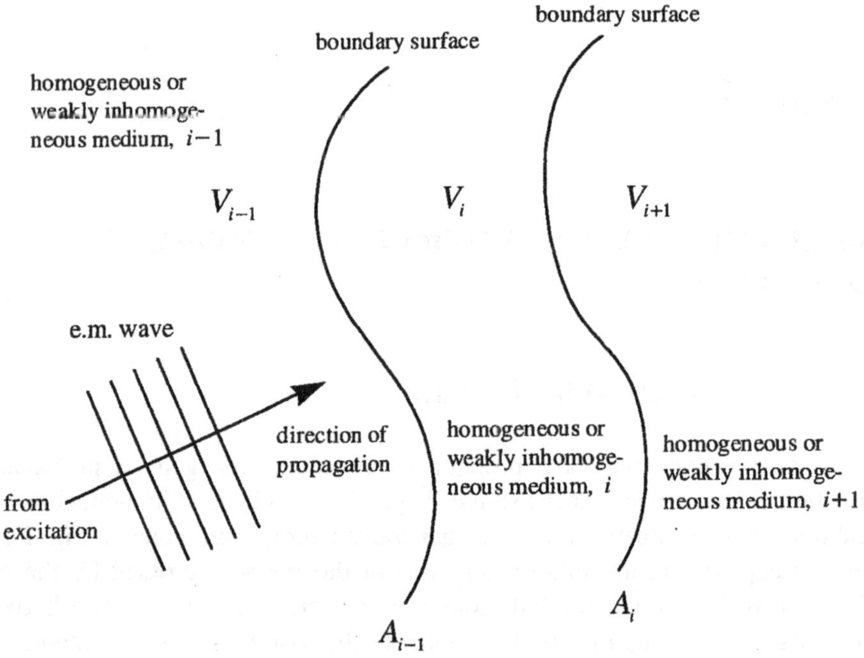

Figure 1.1. The general situation of propagation.

The relations between the different e.m. components could be linear or arbitrary, regular functions

$$\overline{D} \Leftrightarrow (\overline{E}, \overline{H}); \quad \overline{B} \Leftrightarrow (\overline{E}; \overline{H}).$$

Later in this book we shall use *linear* models presenting examples of solutions, in which the medium $\Leftrightarrow$ e.m. signal interaction is *linear*, however, the method is more general. The independent variables are the space vector $\overline{r}(x, y, z)$ and the time t. Describing the structure of medium let us use the notations of Fig. 1.2.

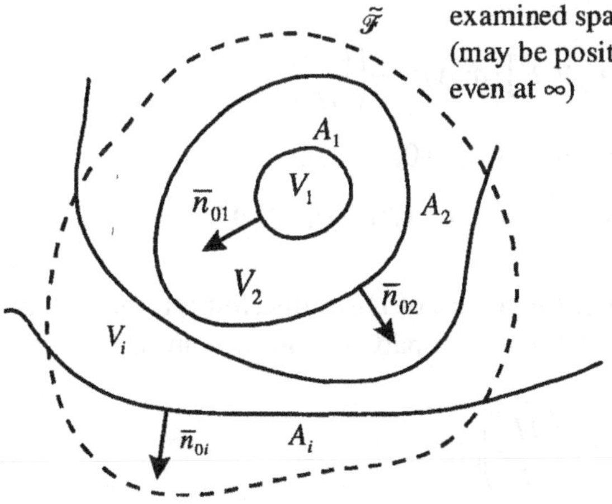

Figure 1.2. The structure of the medium

In the simplest modelling of the presence of distributions let us suppose that (*Idemen*, 1973)

$$
\begin{aligned}
\bar{J}(\bar{r}) &= \sum_{i=1}^{M} \bar{J}_{V_i}(\bar{r}) + \sum_{i=1}^{M} \bar{J}_{A_i}(\bar{r}) \cdot \delta(A_i) = (\bar{J}) + \bar{J}_A \cdot \delta(A), \\
\rho(\bar{r}) &= \sum_{i=1}^{M} \rho_{V_i}(\bar{r}) + \sum_{i=1}^{M} \rho_{A_i}(\bar{r}) \cdot \delta(A_i) = (\rho) + \rho_A \cdot \delta(A),
\end{aligned}
\tag{1.2}
$$

where δ is the Dirac distribution. If a vectorspace $\bar{r}$ contains distributions too, the derivatives of $\bar{u}$ are the following

$$
\begin{aligned}
&\bar{\nabla}\bar{u} = \sum_i \bar{\nabla} \cdot \bar{u}_{V_i} + \sum_i \Delta[\bar{n}_{0i} \times \bar{u}_{A_i}] \cdot \delta(A_i) = (\bar{\nabla} \cdot \bar{u}) + \Delta[\bar{n}_0 \cdot \bar{u}] \cdot \delta(A), \\
&\bar{\nabla} \times \bar{u} = \sum_i \bar{\nabla} \cdot \bar{u}_{V_i} + \sum_i \Delta[\bar{n}_{0i} \times \bar{u}_{A_i}] \cdot \delta(A_i) = (\bar{\nabla} \times \bar{u}) + \Delta[\bar{n}_0 \times \bar{u}] \cdot \delta(A), \\
&\bar{\nabla}(\bar{\nabla} \times \bar{u}) \equiv 0,
\end{aligned}
\tag{1.3}
$$

where $\bar{n}_{0i}$ is the normal unit vector of surfaces A_i and $\Delta[f_i]$ is the step of f at the surface A_i. The surfaces A_i are continuous and do not cross each other. Using (1.3) in (1.1) we get the new form of Maxwell's equation

$$(\overline{\nabla}\times\overline{H})+\Delta[\overline{n}_0\times\overline{H}]\cdot\delta(A)=(J)+\left(\frac{\partial\overline{D}}{\partial t}\right)+\overline{J}_A\cdot\delta(A),$$
$$(\overline{\nabla}\times\overline{E})+\Delta[\overline{n}_0\times\overline{E}]\cdot\delta(A)=-\left(\frac{\partial\overline{B}}{\partial t}\right),\qquad \mathbf{(1.4)}$$
$$(\overline{\nabla}\cdot\overline{B})+\Delta[\overline{n}_0\cdot\overline{B}]\cdot\delta(A)=0,$$
$$(\overline{\nabla}\cdot\overline{D})+\Delta[\overline{n}_0\cdot\overline{D}]\cdot\delta(A)=(\rho)+\rho_A\cdot\delta(A).$$

As a consequence of the nature of Dirac distribution the equation system (1.4) separates to two parts. The first part is valid in volumes V_i

$$(\overline{\nabla}\times\overline{H})=(\overline{J})+\left(\frac{\partial\overline{D}}{\partial t}\right),$$
$$(\overline{\nabla}\times\overline{E})=-\left(\frac{\partial\overline{B}}{\partial t}\right),\qquad \mathbf{(1.5a)}$$
$$(\overline{\nabla}\cdot\overline{B})=0,$$
$$(\overline{\nabla}\cdot\overline{D})=(\rho).$$

In the volumes V_i Maxwell's equations have the well known form and we can use the known methods to get the solutions. The coupling of these solutions at the boundary surfaces A_i is defined by the second part of equation system (1.4), which is valid at surfaces A_i

$$\Delta[\overline{n}_0\times\overline{H}]=\overline{J}_A,$$
$$\Delta[\overline{n}_0\times\overline{E}]=0,\qquad \mathbf{(1.5b)}$$
$$\Delta[\overline{n}_0\cdot\overline{B}]=0,$$
$$\Delta[\overline{n}_0\cdot\overline{D}]=\rho_A.$$

This means that in the presence of distributions the full solution of Maxwell's equations will have steps defined by (1.5b). Therefore if we want to find propagating e.m. wave solutions of Maxwell's equations (*Ferencz,* 1978a) – i.e. e.m. signals propagating across the medium containing distributions according to Fig. 1.1 – it is possible to look for these solutions using the gate functions (distributions) $s_i(\overline{r})$, where

$$s_i(\overline{r})=\{\Theta[\overline{r}(A_{i-1})]-\Theta[\overline{r}(A_i)]\}\qquad \mathbf{(1.6)}$$

and $\Theta(\bar{r})$ is the Heaviside-distribution (unit-jump functional) – see Fig. 1.3.

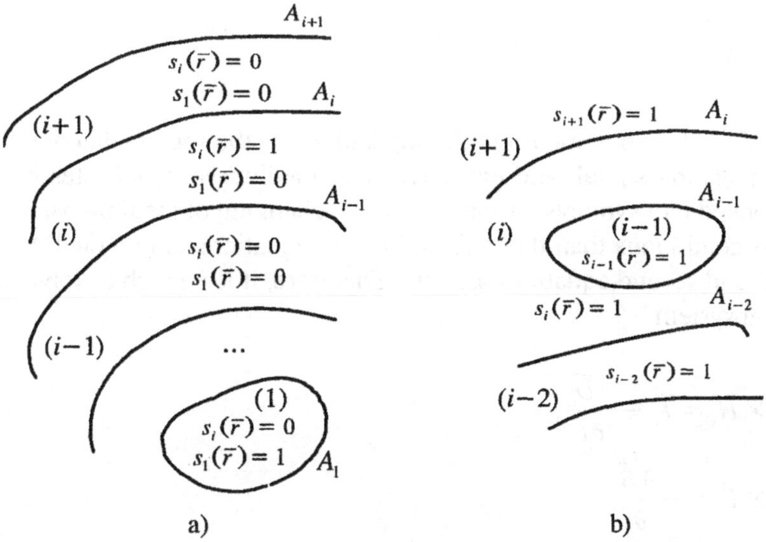

Figure 1.3. The gate functions (distributions) in the cases of different geometrical structures of medium.

Let us find the full wave solutions of Maxwell's equations in the form

$$\bar{G}=\sum_{i=1}^{M}s_i(\bar{r})\left[\sum_{n=1}^{N}\bar{G}_n\right]_i, \tag{1.7}$$

where $\bar{G}$ means $\bar{E},\bar{D},\bar{B}$ and $\bar{H}$ respectively, M is the number of volumes V_i (see Fig. 1.3) and N is the total number of possible solutions (modes) of Maxwell's equations in a volume V_i. If we apply (1.7) in (1.1) the derivative of $s_i(\bar{r})$ will appear there. These derivatives are

$$\bar{\nabla}s_i(\bar{r})=\delta\left[\bar{r}-\bar{r}(A_{i-1})\right]\bar{n}_{0i-1}-\delta\left[\bar{r}-\bar{r}(A_i)\right]\bar{n}_{0i}. \tag{1.8}$$

Applying equations (1.6) ~ (1.8) in (1.1) we can derive the form of equation system in volumes V_i . These equations are in every volume V_i

$$\overline{\nabla}\times\overline{H}_i=\overline{J}_{si}+\frac{\partial\overline{D}_i}{\partial t},$$

$$\overline{\nabla}\times\overline{E}_i=-\frac{\partial\overline{B}_i}{\partial t}, \qquad \textbf{(1.9)}$$

$$\overline{\nabla}\cdot\overline{B}_i=0,$$

$$\overline{\nabla}\cdot\overline{D}_i=\rho_{si}.$$

where $\overline{J}_{si}$ is the source-current density and ρ_{si} is the source-charge density of the propagating signal, and every effect of medium ⇔ signal interactions are incorporated into symbols $\overline{D}_i$ and $\overline{B}_i$. If the changing of medium parameters in space is continuous than the third and fourth equations in (1.9) will be true, if the first and second equations are true. Therefore, it is enough to solve only the equation system

$$\overline{\nabla}\times\overline{H}_i=\overline{J}_{si}+\frac{\partial\overline{D}_i}{\partial t},$$

$$\overline{\nabla}\times\overline{E}_i=-\frac{\partial\overline{B}_i}{\partial t} \qquad \textbf{(1.10)}$$

in every volume V_i respectively.

The full solution is defined by the second part of Maxwell's equations after applying equations (1.6) ~ (1.8), which are valid at the surfaces A_i

$$\sum_{i=1}^{M}\overline{\nabla}s_i(\overline{r})\times\left[\sum_{n=1}^{N}\overline{H}_n\right]_i=0,$$

$$\sum_{i=1}^{M}\overline{\nabla}s_i(\overline{r})\times\left[\sum_{n=1}^{N}\overline{E}_n\right]_i=0,$$

$$\sum_{i=1}^{M}\overline{\nabla}s_i(\overline{r})\cdot\left[\sum_{n=1}^{N}\overline{B}_n\right]_i=0, \qquad \textbf{(1.11)}$$

$$\sum_{i=1}^{M}\overline{\nabla}s_i(\overline{r})\cdot\left[\sum_{n=1}^{N}\overline{D}_n\right]_i=0.$$

It is important to remark that the derivation of the signal ⇔ medium interactions is not too simple, because this interaction depends on the real shape of the signal itself even in linear cases. For example the result of this interaction – permittivity, permeability or more general relations – is essentially different if the signal is a monochromatic wave or a monochromatic wave with elementary small perturbation or other ones (*Ferencz*, 1978b).

Another problem is the applied mathematical way finding the solution of this partial differential equation-system. In general, many authors use the Fourier-transformation of equation-system (1.9) if the shape of excitation is arbitrary. After the Fourier transformation they apply in the transformed equation system the constitutional relations (permittivity, permeability etc.) of monochromatic signals using the connections $\overline{D} \Leftrightarrow \overline{E}$ or $\overline{D} \Leftrightarrow (\overline{E}, \overline{H})$ and $\overline{B} \Leftrightarrow \overline{H}$ or $\overline{B} \Leftrightarrow (\overline{E}, \overline{H})$ of monochromatic cases. However, in this mathematical method two highly problematic points exist. The first problem is the nature of Fourier-transformation itself. The Fourier-transformation operates with the orthogonal trigonometric function-space. These (orthogonal) functions exist from $\xi = -\infty$ to $\xi = +\infty$, where ξ is the independent variable, as the time t or the space-vector components (x, y, z). Therefore this method has a well known consequence that every signal described by this manner is "predestinated" just now at $\xi = -\infty$. However, in real cases the appearing signals (switched-in signals) have not this character. Therefore in network theory we do not use the Fourier-transformation in real transient investigations. The application of Laplace-transformation is more convenient. The second problem is the finding of the constitutional relations. Because these relations essentially depend on the shape of signals (*Ferencz*, 1978b), using the constitutional relations of monochromatic signals is inaccurate. To find a good solution we must go back to the basic relations defining the elementary interaction between the particles of the medium and the e.m. signal in every case deriving a real solution generated by arbitrary shaped excitation.

1.2 Laplace-transformation with multi variables

The well known definition of Laplace-transformation is

$$\mathscr{L}[f(t)] \equiv F(s) = \int_0^\infty f(t) e^{-st} dt, \tag{1.12}$$

where t and s are the conjugated independent variables $(t \Leftrightarrow s), f(t)$ is a function with real or complex values, t is an independent variable with real values and s is an independent variable with complex values (i.e. $s = \sigma + j\omega$) and the integral in convergent, if $\mathrm{Re}\, s = \sigma > \sigma_a$, where σ_a is a constant. In this case the values of $f(t)$ function in the $-\infty < t < 0$ half-space are definitely zero (or, in other words, the values of $f(t)$ in the half-space $t < \infty$ have no meaning). The place of the origo $t = 0$ is a question of definition knowing the real process in our investigations. The definition of inverse transformation is the Fourier-Mellin formula

$$\mathscr{L}^{-1}[F(s)] \equiv f(t) = \frac{1}{2\pi \mathrm{j}} \int_{c-\mathrm{j}\infty}^{c+\mathrm{j}\infty} F(s) \mathrm{e}^{st} \mathrm{d}s, \tag{1.13}$$

where c is a constant and $c > \sigma_a$. However, in practice we use known rules to solve the task of inverse transformation (see e.g. *Simonyi*, 1963) and we suppose that the reader knows the Laplace-transforms of different known functions and functionals.

The wave propagation tasks are mathematical problems with multi variables. Because the generalisation of (1.12) is not generally known, let us see the most frequently used relations in these cases.

Single variable case:

$$\begin{aligned} \mathscr{L}\left[\frac{\mathrm{d}f}{\mathrm{d}t}\right] &= \int_0^\infty \frac{\mathrm{d}f(t)}{\mathrm{d}t} \mathrm{e}^{-st} \mathrm{d}t = \left[f(t)\mathrm{e}^{-st}\right]_{t=0}^{\infty} + s\int_0^\infty f(t)\mathrm{e}^{-st} \mathrm{d}t = \\ &= sF(s) - f(t=0) = sF(s) - f_0 \end{aligned} \tag{1.14}$$

and

$$\begin{aligned} \mathscr{L}\left[\frac{\mathrm{d}^2 f}{\mathrm{d}t^2}\right] &= \int_0^\infty \frac{\mathrm{d}^2 f(t)}{\mathrm{d}t^2} \mathrm{e}^{-st} \mathrm{d}t = \left[\frac{\mathrm{d}f(t)}{\mathrm{d}t}\mathrm{e}^{-st}\right]_{t=0}^{\infty} + \int_0^\infty \frac{\mathrm{d}f(t)}{\mathrm{d}t} \mathrm{e}^{-st} \mathrm{d}t = \\ &= s[sF(s) - f(t=0)] - \left.\frac{\mathrm{d}f}{\mathrm{d}t}\right|_{t=0} = s^2 F(s) - sf_0 - f_0' \end{aligned} \tag{1.15}$$

where $f_0 = f(t=0)$ and $f_0 = \left.\frac{df(t)}{dt}\right|_{t=0}$ are the boundary conditions of the problem.

Let the conjugated variables in the *multi variable case* be:

$$t \Leftrightarrow s, \quad x \Leftrightarrow p. \tag{1.16}$$

The most important relations are

$$\mathscr{L}[f(x,t)] \equiv F(p,s) = \int_0^\infty\!\!\int f(x,t) e^{-px} e^{-st} \mathrm{d}x\mathrm{d}t. \tag{1.17}$$

Therefore

$$\begin{aligned}
\mathscr{L}\left[\frac{\partial f(x,t)}{\partial t}\right] &= \int_0^\infty\!\!\int \frac{\partial f(x,t)}{\partial t} e^{-st} e^{-px} \mathrm{d}t = \\
&= \int_0^\infty \left\{ \int_0^\infty \frac{\partial f(x,t)}{\partial t} e^{-st} \mathrm{d}t \right\} e^{-px} \mathrm{d}x = \\
&= \int_0^\infty \left\{ s\int_0^\infty f(x,t) e^{-st} \mathrm{d}t - f(x,t=0) \right\} e^{-px} \mathrm{d}x = \\
&= s\int_0^\infty\!\!\int f(x,t) e^{-px} e^{-st} \mathrm{d}x\,\mathrm{d}t - \int_0^\infty f_{x0}(x) e^{-px} \mathrm{d}x = sF(p,s) - f_{x0}(p),
\end{aligned} \tag{1.18}$$

where $f_{x0}(p)$ is the Laplace transform of the boundary condition $f(x,t=0) \triangleq f_{x0}(x) = f_{x0}$. (Using these notations the $f(x,t)$ function could be written as f_{xt}.)

$$\begin{aligned}
\mathscr{L}\left[\frac{\partial^2 f(x,t)}{\partial t^2}\right] &= \int_0^\infty\!\!\int \frac{\partial^2 f(x,t)}{\partial t^2} e^{-st} e^{-px} \mathrm{d}t\,\mathrm{d}x = \\
&= \int_0^\infty \left\{ \int_0^\infty \frac{\partial^2 f(x,t)}{\partial t^2} e^{-st} \mathrm{d}t \right\} e^{-px} \mathrm{d}x = \\
&= \int_0^\infty \left\{ s\int_0^\infty \frac{\partial f(x,t)}{\partial t} e^{-st} \mathrm{d}t - f'_{x0}(x) \right\} e^{-px} \mathrm{d}x = \\
&= s^2 F(p,s) - s f_{x0}(p) - f'_{x0}(p),
\end{aligned} \tag{1.19}$$

where $f'_{x0}(x) = \left.\dfrac{\partial f(x,t)}{\partial t}\right|_{t=0}$ and $f'_{x0}(p)$ is the Laplace transform of $f'_{x0}(x)$.

$$
\begin{aligned}
&\mathcal{L}\left[\frac{\partial^2 f(x,t)}{\partial t}\right] = \int_0^\infty\!\!\int \frac{\partial^2 f(x,t)}{\partial x\, \partial t} \mathrm{e}^{-px}\mathrm{e}^{-st}\mathrm{d}x\,\mathrm{d}t = \\
&= \int_0^\infty \left\{ s\int_0^\infty \frac{\partial^2 f(x,t)}{\partial x \partial t}\mathrm{e}^{-st}\mathrm{d}t \right\}\mathrm{e}^{-px}\mathrm{d}x = \int_0^\infty \left\{ \int_0^\infty \frac{\partial}{\partial t}\left[\frac{\partial f(x,t)}{\partial x}\right]\mathrm{e}^{-st}\mathrm{d}t \right\}\mathrm{e}^{-px}\mathrm{d}x = \\
&= \int_0^\infty \left\{ s\int_0^\infty \frac{\partial f(x,t)}{\partial x}\mathrm{e}^{-st}\mathrm{d}t \right\}\mathrm{e}^{-px}\mathrm{d}x - \int_0^\infty \left.\frac{\partial f(x,t)}{\partial x}\right|_{t=0} \mathrm{e}^{-px}\mathrm{d}x = \\
&= \int_0^\infty \left\{ \int_0^\infty \frac{\partial f(x,t)}{\partial x}\mathrm{e}^{-px}\mathrm{d}x \right\}\mathrm{e}^{-st}\mathrm{d}t - \int_0^\infty \left.\frac{\partial f(x,t)}{\partial x}\right|_{t=0} \mathrm{e}^{-px}\mathrm{d}x = \\
&= psF(p,s) - s\int_0^\infty f(x=0,t)\mathrm{e}^{-st}\mathrm{d}t - \int_0^\infty \left.\frac{\partial f(x,t)}{\partial x}\right|_{t=0} \mathrm{e}^{-px}\mathrm{d}x = \\
&= psF(p,s) - sf_{0t}(s) - pf_{x0}(p) + f_{00},
\end{aligned}
\tag{1.20}
$$

where $f_{00} = f(x=0, t=0)$ and $f_{0t}(s)$ is the Laplace transform of function $f(x=0,t)$.

In the following, it is supposed that the commonly used basic relations of Laplace transformation (e.g. Laplace and inverse-Laplace transforms of trigonometric functions, of Dirac and Heaviside distributions etc. or the relation between the Laplace and Fourier transforms in the case of $s = j\omega$ and others) are known by the readers.

1.3 Application of the method in a homogeneous magnetoionic medium for longitudinal propagation

One of the deeply investigated phenomenon between the non-monochromatic, general shaped (i.e. extra wide band) signals is the whistler-propagation in the high atmosphere of the Earth and of other planets having their own magnetic fields. Whistlers are electromagnetic signals travelling across a magnetised plasma along magnetic field lines or in oblique directions. (See Fig. 1.4 as an example of longitudinal whistler propagation along geomagnetic field lines.) These signals are generated by lightning in the troposphere and have great dispersion (Fig. 1.5). The whistlers might as well have multiple reflection between the hemispheres.

The theoretical modelling of the longitudinal propagation of *monochromatic* waves in a magnetised (anisotropic) plasma has shown, as is known (*Storey*, 1953), that below the gyrofrequency (ω_b) the group velocity (v_g) of the signal exhibits a characteristic dispersion. Accordingly, signals generated for example by lightning enter the magnetosphere and propagating in it (in a quasi-longitudinal mode) suffer dispersion. The resulting signal (whistler)

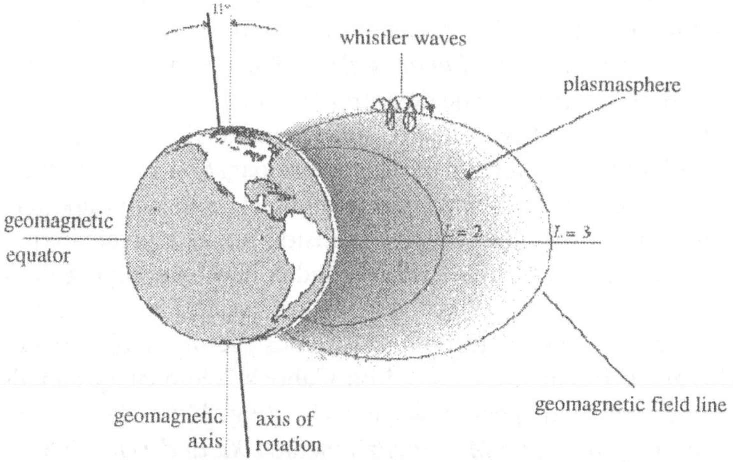

Figure 1.4. Propagation of a whistler in the magnetosphere. (*L* means the equatorial distance of a geomagnetic field line from the centre of the Earth in the geomagnetic equatorial plane, measuring this distance in Earth's radius).

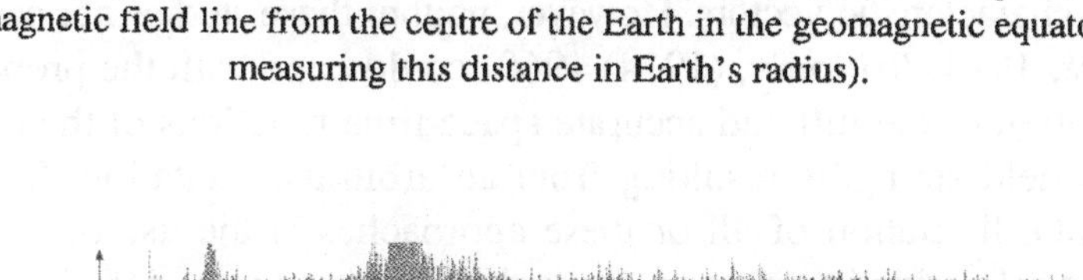

Figure 1.5. Time function and dynamic spectrum of a measured whistler (No. TS84010 recorded at Halley Station, Antarctica on 21 July 1984, 17:14:14 UT. – Courtesy of dr. A.J. Smith, British Antarctic Survey, Cambridge, U.K.)

propagates along the magnetic lines of force (ducted presumably by ducts), and exhibits a frequency-time function in accordance with the frequency dependence of v_g. In describing the shape of the signal, either the earlier well-known models were applied which use the results of the monochromatic solution mentioned above (e.g. *Bernard*, 1973; *Hamar et al.*, 1990) or various approximations were used (e.g. *Budden*, 1968). The determination of the actual signal form (i.e. the space-time function of the electric and magnetic field strength components) under conditions characteristic of whistler propagation attracted the attention of a number of researchers. There are three characteristic and important classes of approximations. Using the Taylor expansion about a given frequency of the frequency-dependent plasma refractive index and by applying the Airy phase Budden derived an approximate waveform around the nose of a whistler (*Budden*, 1968). Another technique for constructing approximate waveform around a given frequency is based on Gabor's "elementary signals" (*Gabor*, 1946) to construct a dispersive wave packet (e.g. *Mihovilovic and Bracewell*, 1992). Recently, based on *Marcuwitz's* method *Xu and Yeh* (1990) constructed a VLF electromagnetic wave packet by superposing a group of spectrum-weighted characteristic vectors. However, neither these studies nor earlier works (*Havelock*, 1914, *Brillouin*, (1914) 1960) could cope with the problem of the determination of the full and accurate space-time functions of the electric and magnetic field strengths resulting from an arbitrary excitation. In general, a fundamental limitation of all of these approaches is the use of a frequency-dependent refractive index or an equivalent quantity (e.g. permittivity tensor, wave packet approximation) which assumes at the beginning of the derivations an interaction between a monochromatic signal and plasma. However, without incorporating the signal source it is impossible to get a full and accurate solution for the space-time functions of field strength components. It is to be noted, that the starting formula used by *Xu and Yeh* (1990) is, in fact, non-monochromatic and satisfactorily general, but later on they return to the quasi-monochromatic wave packet approximation. In addition to the above-mentioned approaches, there are a number of papers treating the whistler wave propagation in homogeneous or inhomogeneous magnetoionic media. All of these use a monochromatic or quasi-monochromatic model for the propagating wave field, and are aimed at computing the direction of propagation, self focusing, ducted propagation and instabilities (e.g. *Karpman and Kaufman*, 1981; 1982a,b, 1987; *Karpman and Shagalov*, 1989; *Laird*, 1992), or the propagation of whistler wave packets from the point of view of propagation velocity (e.g. *Suchy*, 1972; *Anderson et al.*, 1975, 1976; *Tanaka et al.*, 1986). According to their purposes, none of these papers present a rigorous (or even approximate) description of the spatial and temporal function of the field strength (i.e. the waveform), which would be suitable for the modelling of actually measured whistler waveforms, recorded on board of spacecrafts or at the ground. It was stressed by *Laird*

(1992) that to answer some delicate questions related to the fine structure and propagation of whistlers "one probably needs some form of full wave theory". In this section it is demonstrated that a real full wave solution can be derived using the method from point 1.1 (MIBM) and it has been applied successfully for propagation research (*Ferencz,* 1995a).

1.3.1 Formulation of the problem

Let us use a simplified version of the model sketched on Fig.1.3 – see Fig.1.6. In this model, the atmospheric layers are modelled in Cartesian coordinate-system and the excitation appears in the half-space "1". Medium "1" is the Earth's surface-ionospheric wave guide, further it is called "lower atmosphere", where the arbitrarily chosen non-monochromatic exciting current density generated by the lightning discharge impulse arises. It is important to emphasise that there is not any "traditional" wavepacket model or other monochromatic approximations. Medium "1" is modelled in these investigations by vacuum, as usual. In this half-space the medium has no significant interaction to the signal, i.e. the medium is homogeneous, isotropic, time-invariant and characterised by (ε_0, μ_0) Therefore, we can concentrate on the role of the magnetoionic medium and of the extra wide band excitation. The excitation signal appears inside of half-space "1" (volume V_1). Let the excitation start in the $0 \le t \le t_0$ time interval and in the $0 \le x \le x_*$ space interval, where $x_* \le x_0$. Let x_0 be the place of the boundary surface (i.e. boundary plane $x = x_0$) between the volumes V_1 and V_2 (half-spaces "1" and "2"). Let the direction of excitation, i.e. the direction of source current density be parallel to the z-axis, i.e.

$$\overline{J}_1(x,t) = J_0(x,t)\overline{e}_z . \tag{1.21}$$

In the half-space "2" (i.e. $x \le x_0$) a magnetoionic medium is present in which the static magnetic field (the superimposing geomagnetic field) is parallel to x-axis, i.e.

$$\overline{B}_{F0} = B_{F0}\,\overline{e}_x , \tag{1.22}$$

where $\overline{B}_{F0}$ = constant in a homogeneous case. The medium "2" – i.e. the magnetosphere – is supposed to be an infinite half space (the phenomenon of the so called multi-hop whistlers – the signal arriving from the magnetosphere is reflected from the boundary surface – will not be dealt with in this simple case).

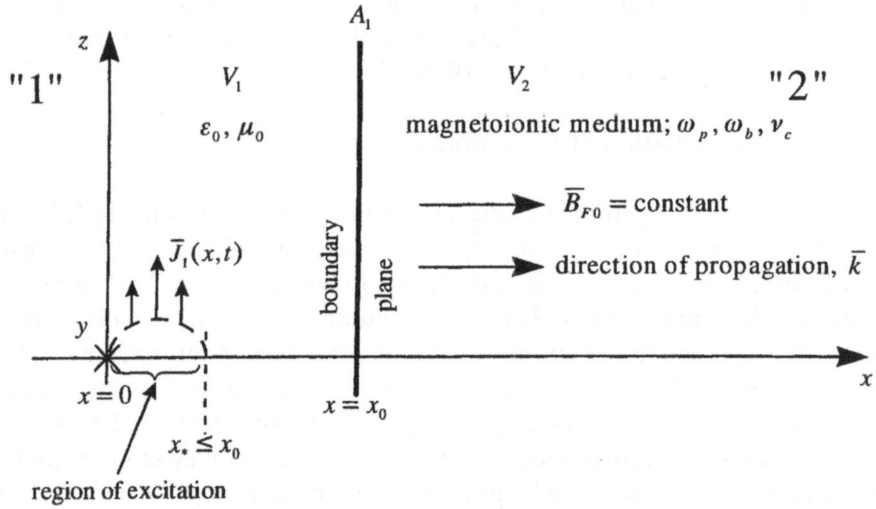

Figure 1.6. Model for longitudinal propagation; the medium in half-space "1" is homogenous and in half-space "2" is homogeneous or inhomogeneous depending on the given problem.

The boundary plane at $x = x_0$ is perpendicular to *x*-axis, i.e.

$$\overline{n}_0 = \overline{e}_x. \qquad \textbf{(1.23)}$$

In this case, we wish now to find the e.m. signal (wave) propagating in half-space "2" and generated by an excitation appearing in half-space "1". The main steps of the solution will be in accordance with the theoretical background (point 1.1): derivation of the existing modes in half-space "2"; derivation of the existing modes in half-space "1"; solution of equation-system (1.11) at the boundary-plane A_i.

1.3.2 The propagation problem in half-space "2"

First, let us examine the determination of the complete propagating signal in a homogeneous, time independent, linear, cold, magnetised, single component (i.e. electron) plasma. In this model we ignore the motion of ions, the loss effects (*O. Ferencz*, 1993) and thermal effects (*Sazhin*, 1993), from which the loss and ion motions will be taken into account in Chapter 2. The generalisation for an inhomogeneous plasma and for non-longitudinal (oblique) propagation will be discussed also in Chapter 2.

The non-monochromatic, propagating electromagnetic signal of unknown shape is defined by the actual form of equation (1.10), i.e.

$$\nabla \times \overline{H}_2 = \overline{J} + \varepsilon_0 \frac{\partial \overline{E}_2}{\partial t},$$
$$\nabla \times \overline{E}_2 = -\mu_0 \frac{\partial \overline{H}_2}{\partial t}, \tag{1.24}$$

where ε_0 and μ_0 are the permittivity and permeability of vacuum, respectively, and $\overline{J}$ is generated by the e.m. signal $(\overline{E},\overline{H})$ itself in the magnetoionic medium. In the given homogeneous, single-component plasma characterised by an electron density of N, and in the presence of a superimposed magnetic field $\overline{B}_{F0}$, the following (phenomenological) equations describe the signal-medium interaction

$$m\frac{\partial \overline{v}}{\partial t} = q(\overline{E} + \overline{v} \times \overline{B}_{F0}),$$
$$\overline{J} = qN\overline{v}, \tag{1.25}$$
$$\nabla \cdot \overline{J} + \frac{\partial \rho}{\partial t} = 0,$$

where m and $q = -e$ are the mass and charge of an electron, respectively, and $\overline{v}$ is the speed of electrons, due to $\overline{E}$ in the given conditions.

Introducing the

$$\overline{\overline{I}}_B = \begin{vmatrix} 0 & 0 & 0 \\ 0 & 0 & 1 \\ 0 & -1 & 0 \end{vmatrix}$$

tensor, the first equation of (1.25), which describes the interaction between the signal and the medium, can be written in the following, more convenient form

$$\frac{\partial \overline{v}}{\partial t} + \omega_b \overline{\overline{I}}_B \overline{v} = \frac{q}{m} \overline{E}(\overline{r},t), \tag{1.26}$$

where

$$\omega_b = \frac{eB_{F0}}{m}$$

is the angular electron gyrofrequency. The equation (1.26) is a known type of differential equations and can be solved even if we do not know the form of $\overline{E}$, because we have to solve the following system of equations

$$\frac{\partial \overline{v}}{\partial t}+\overline{\overline{a}}\cdot\overline{v}=\overline{f}(\overline{r},t), \tag{1.26a}$$

i.e. in component form

$$\begin{aligned}
&\frac{\partial v_x}{\partial t}=f_x(\overline{r},t),\\
&\frac{\partial v_y}{\partial t}+a\cdot v_z=f_y(\overline{r},t),\\
&\frac{\partial v_z}{\partial t}-a\cdot v_y=f_z(\overline{r},t).
\end{aligned} \tag{1.26b}$$

Therefore the equation (1.26) can be reduced to an ordinary system of differential equations (*Kamke*, 1965) on the one hand, and on the other hand it can be converted into three differential equations of the same type, which have closed form solutions (*Kamke*, 1956). If we use the notation $\partial u/\partial t=u'$ and $\partial^2 u/\partial^2 t=u''$ so we can write (1.26b) in the form

$$\begin{aligned}
&v_x'=f_x,\\
&v_y'+a\cdot v_z=f_y,\\
&v_z'-a\cdot v_y=f_z.
\end{aligned} \tag{1.26c}$$

The solution of the first equation in (1.26c) is trivial. The second and third equations after trivial conversions

$$\begin{aligned}
&v_y''+a^2\cdot v_y=f_y'+a\cdot f_z,\\
&v_z''+a^2\cdot v_z=f_z'+a\cdot f_z.
\end{aligned} \tag{1.26d}$$

These equations have the same mathematical type, which is

$$y''+\alpha y'+\beta y=\varphi(x),$$

using the notations of Kamke. In this case $\alpha\equiv 0$. Based on known considerations, we can neglect the homogeneous solutions of these differential equations, because these are independent of any excitation, i.e. in this case on $\overline{E}(\overline{r},t)$. The inhomogeneous solutions corresponding to the excitation $\overline{E}(\overline{r},t)$ are the following (see e.g. *Kamke*, 1956)

$$v_x = \frac{q}{m}\int^t E_x(\overline{r},\tau)\cdot \mathrm{d}\tau,$$

$$v_y = C_y\left\{\frac{q}{m\omega_b}\int^t \frac{\partial E_y(\overline{r},\tau)}{\partial \tau}\sin\omega_b(t-\tau)\cdot \mathrm{d}\tau - \frac{q}{m}\int^t E_z(\overline{r},\tau)\sin\omega_b(t-\tau)\cdot \mathrm{d}\tau\right\},$$

$$v_z = C_z\left\{\frac{q}{m\omega_b}\int^t \frac{\partial E_z(\overline{r},\tau)}{\partial \tau}\sin\omega_b(t-\tau)\cdot \mathrm{d}\tau - \frac{q}{m}\int^t E_y(\overline{r},\tau)\sin\omega_b(t-\tau)\cdot \mathrm{d}\tau\right\},$$

(1.27)

where C_y and C_z are constants. Applying v_y and v_z in equation (1.26) – as these are given in (1.27) – and making some manipulations to simplify the relations, it is possible to determine the C_y and C_z constants. The result is (see Appendix A)

$$C_y = C_z = 1. \qquad \textbf{(1.27a)}$$

By applying (1.27) in the second equation of (1.25) we can determine the current density $\overline{J} = qN\overline{v}$ generated by the $\overline{E}(\overline{r},t)$ signal in the magnetoionized medium "2". If we define the angular plasma frequency (ω_p) of electrons as

$$\omega_p^2 = \frac{q^2 N}{\varepsilon_0 m} \equiv \frac{e^2 N}{\varepsilon_0 m}, \qquad \textbf{(1.28a)}$$

the components of current density are given as

$$J_x = \varepsilon_0\omega_p^2\int^t E_x \mathrm{d}\tau,$$

$$J_y = \varepsilon_0\omega_p^2\left\{\frac{1}{\omega_b}\int^t \frac{\partial E_y}{\partial \tau}\sin\omega_b(t-\tau)\mathrm{d}\tau - \int^t E_z\sin\omega_b(t-\tau)\mathrm{d}\tau\right\}, \qquad \textbf{(1.28)}$$

$$J_z = \varepsilon_0\omega_p^2\left\{\frac{1}{\omega_b}\int^t \frac{\partial E_z}{\partial \tau}\sin\omega_b(t-\tau)\mathrm{d}\tau + \int^t E_y\sin\omega_b(t-\tau)\mathrm{d}\tau\right\}.$$

Applying Eq. (1.28) we can show – as a good control of our first steps as well – that

• supposing monochromatic $(\overline{E},\overline{H})$ signal we get the well known monochromatic solutions (see Appendix B);

• in case of a really arbitrary shape $(\overline{E},\overline{H})$ signal, e.g. in the case of a Dirac delta shape excitation, the definition of the wave crests is impossible. For example, a Dirac delta signal has only one wave crest (peak – with infinite amplitude), however, in this signal the possibility of an infinite number of wave crests are present. Therefore, the fulfilment of the conservation of wave crests (see e.g. *Ferencz*, 1996, pp. 267-268.) has no meaning in these cases. However, if during the propagation of signal the shape of it could be written sectional in the form

$$\overline{G} \sim \overline{G}_0 e^{j\varphi},$$

where $\overline{G}$ means $\overline{E},\overline{D},\overline{H},\overline{B}$ and $\overline{G}_0$ is the amplitude (a real value) and

$$\phi = \int^{t} \Omega(x,y,z,\tau)\mathrm{d}\tau - \int^{x} K_x(\xi,y,z,t)\mathrm{d}\xi -$$
$$- \int^{y} K_y(x,\eta,z,t)\mathrm{d}\eta - \int^{z} K_z(x,y,\zeta,t)\mathrm{d}\zeta \ ,$$

than the conservation of wave crest will always fulfil (see Appendix C).

If we write equations (1.28) back into Maxwell's equations (1.24), then the system of equations to be solved is

$$\frac{\partial H_z}{\partial y} - \frac{\partial H_y}{\partial z} = \varepsilon_0 \cdot \left\{ \omega_p^2 \int^t E_x \mathrm{d}\tau + \frac{\partial E_x}{\partial t} \right\},$$
$$\frac{\partial H_x}{\partial z} - \frac{\partial H_z}{\partial x} = \varepsilon_0 \cdot \left\{ \frac{\omega_p^2}{\omega_b} \int^t \frac{\partial E_y}{\partial \tau} \sin\omega_b(t-\tau)\mathrm{d}\tau - \omega_p^2 \int^t E_z \sin\omega_b(t-\tau) \cdot \mathrm{d}\tau + \frac{\partial E_y}{\partial t} \right\},$$
$$\frac{\partial H_y}{\partial x} - \frac{\partial H_x}{\partial y} = \varepsilon_0 \cdot \left\{ \frac{\omega_p^2}{\omega_b} \int^t \frac{\partial E_z}{\partial \tau} \sin\omega_b(t-\tau)\mathrm{d}\tau + \omega_p^2 \int^t E_y \sin\omega_b(t-\tau) \cdot \mathrm{d}\tau + \frac{\partial E_z}{\partial t} \right\}, \ \textit{cont}$$

(1.29)

$$\frac{\partial E_z}{\partial y}-\frac{\partial E_y}{\partial z}=-\mu_0\frac{\partial H_x}{\partial t},$$

$$\frac{\partial E_x}{\partial z}-\frac{\partial E_z}{\partial x}=-\mu_0\frac{\partial H_y}{\partial t}, \qquad \textbf{(1.29)}$$

$$\frac{\partial E_y}{\partial x}-\frac{\partial E_x}{\partial y}=-\mu_0\frac{\partial H_z}{\partial t}.$$

In agreement with our goal, we want to find only that solution of (1.29) which describes a strictly "longitudinal" plane wave propagating in the $\overline{e}_x$ direction. (Of course, the method can also be used in studying quasi longitudinal signals, and signals propagating in arbitrary directions – see point 2.5 –, and can be generalised – with mathematical complexities – for non-planar wave models too.) Let us transform equations (1.29), in a few steps, into three equations including only E_i components applying the conditions of the model $\partial/\partial z=\partial/\partial y=0$. I.e. let us multiply the first three equations in (1.29) by $\partial/\partial t$. After this, the second three equations of (1.29) and the $\partial/\partial y=\partial/\partial z=0$ can be used in the first three ones. The operations $\int f(t,\tau)\mathrm{d}\tau$ and $\partial/\partial t$ are interchangeable. After these manipulations, the first equation reads as

$$\frac{1}{c^2}\left(\omega_p^2E_x+\frac{\partial^2E_x}{\partial t^2}\right)=0, \qquad \textbf{(1.30a)}$$

where c is the speed of light in free space and $c^2=1/\varepsilon_0\mu_0$. This equation gives $E_x=0$. We still have to solve the other two equations

$$\frac{\partial E_y}{\partial x^2}=\frac{1}{c^2}\left\{\omega_p^2\int_0^t\frac{\partial E_y}{\partial\tau}\cos\omega_b(t-\tau)\cdot\mathrm{d}\tau-\omega_b\omega_p^2\int_0^t E_z\cos\omega_b(t-\tau)\cdot\mathrm{d}\tau+\frac{\partial^2E_y}{\partial t^2}\right\},$$

$$\textbf{(1.30b)}$$

$$\frac{\partial E_z}{\partial x^2}=\frac{1}{c^2}\left\{\omega_p^2\int_0^t\frac{\partial E_z}{\partial\tau}\cos\omega_b(t-\tau)\cdot\mathrm{d}\tau+\omega_b\omega_p^2\int_0^t E_y\cos\omega_b(t-\tau)\cdot\mathrm{d}\tau+\frac{\partial^2E_y}{\partial t^2}\right\}.$$

where the chosen starting point is $t=0$.

According to equations (1.30), in the general non-monochromatic case, for strictly longitudinal propagation the signal is transverse just like in the monochromatic case.

1.3.3 Longitudinal plane wave solution in half-space "2"

In this homogeneous case not only c but also ω_p and ω_b are constants, therefore the multi-variable – in this case two-variable – Laplace-transform of equation (1.30b) can be computed. Let us define the complementer variables as

$$x \Leftrightarrow p, \quad t \Leftrightarrow s, \tag{1.31}$$

as was presented in part 1.2.

The derivatives figuring in (1.30b) also require the specification of the boundary condition functions. These and their Laplace-transforms are given as

$$\begin{aligned}
E_i(x,t=0) = e_{ix0}(x) \quad &\Leftrightarrow \quad e_{ix0}(p),\\
E_i(x=0,t) = e_{i0t}(t) \quad &\Leftrightarrow \quad e_{i0t}(s),\\
\left.\frac{\partial E_i(x,t)}{\partial x}\right|_{x=0} = e'_{i0t}(t) \quad &\Leftrightarrow \quad e'_{i0t}(s),\\
\left.\frac{\partial E_i(x,t)}{\partial t}\right|_{t=0} = e'_{ix0}(x) \quad &\Leftrightarrow \quad e'_{ix0}(p),
\end{aligned} \tag{1.32}$$

where $i = y, z$. Without imposing restrictions on the generality of the solution, the relations (1.32) can be further specified. First of all, let the coordinate axes y and z be chosen so that the boundary condition functions in the direction of y $(i = y)$ have a zero value, provided that the excitation is plane-polarised at the plasma boundary, in this example.

This is a feasible approximation if we recall that the source of the excitation signal is a lightning discharge. In addition, we do not place any constraint on the generality of the solution if the time origin $t = 0$ is chosen so that at this time still there is not any signal in the plasma, that is

$$\begin{aligned}
&e_{yx0}(x) \equiv e_{y0t}(t) \equiv e'_{yx0}(x) \equiv e'_{y0t}(t) \equiv 0 \text{ and}\\
&e_{zx0} \equiv 0.
\end{aligned} \tag{1.33}$$

Using (1.31), (1.32) and (1.33) we can obtain the Laplace-transform of equations (1.30b)

$$c^2p^2E_y(p,s)=\omega_p^2\frac{s^2}{s^2+\omega_b^2}E_y(p,s)-\omega_b\omega_p^2\frac{s}{s^2+\omega_b^2}E_z(p,s)+s^2E_y(p,s),$$

$$c^2p^2E_z(p,s)=\omega_p^2\frac{s^2}{s^2+\omega_b^2}E_z(p,s)+\omega_b\omega_p^2\frac{s}{s^2+\omega_b^2}E_y(p,s)+s^2E_z(p,s)+$$
$$+c^2[pe_{z0t}(s)+e'_{z0t}(s)]-e'_{zx0}(p),$$

(1.34)

If the wave propagation is treated in a plasma of infinite extension $[x_0,+\infty]$ then the condition $e'_{zx0}(x)=0$ always holds, while for a signal entering the plasma from another medium ("1") in time-interval $t\geq 0$ the appearing of the signal will not start at the moment $t=0$. Now we assume $e'_{zx0}(p)=0$ which does not represent any specific constraint. (If $e'_{zx0}(p)$ exists in more complicated situations, then it can be approximated by a polinom of p with a degree of not greater than 4.)

From equation (1.34) the field intensity components

$$E_y(p,s)=c^2\frac{a_\infty(s)a_0(s)a_2(s)}{b_3(s)}\cdot\frac{p+\dfrac{a_1(s)}{a_2(s)}}{p^4-\dfrac{b_2(s)}{b_3(s)}\cdot p^2+\dfrac{b_1(s)}{b_3(s)}},$$

(1.35)

$$E_z(p,s)=c^2\frac{a_\infty(s)a_4(s)a_2(s)}{b_3(s)}\cdot\frac{\left[p^2-\dfrac{a_3(s)}{a_4(s)}\right]\left[p+\dfrac{a_1(s)}{a_2(s)}\right]}{p^4-\dfrac{b_2(s)}{b_3(s)}\cdot p^2+\dfrac{b_1(s)}{b_3(s)}},$$

where

$a_\infty(s)=s^2+\omega_b^2;$ $\quad a_2(s)=e_{z0t}(s);$

$a_0(s)=-\omega_b\omega_p^2 s;$ $\quad a_3(s)=s^2(\omega_p^2+\omega_b^2+s^2);$

$a_1(s)=e'_{z0t}(s);$ $\quad a_4(s)=c^2a_\infty(s);$

$b_1(s)=(\omega_b\omega_p^2 s)^2+s^4(\omega_p^2+\omega_b^2+s^2)^2;$

$b_2(s)=2c^2s^2(s^2+\omega_b^2)(\omega_p^2+\omega_b^2+s^2)^2;$

$b_3(s)=c^4(s^2+\omega_b^2)^2.$

It is clear from (1.35) that the inverse Laplace-transform of the field intensity components with respect to place can be computed. (Even if $e'_{zx0}(p) \neq 0$.) The denominator in (1.35) is a polinom of fourth degree with the following roots

$$p_{1,2,3,4} = \pm \frac{1}{c} \sqrt{\frac{s^2\left(s^2+\omega_p^2+\omega_b^2\right) \pm j\left(\omega_b \omega_p^2 s\right)}{s^2+\omega_b^2}} . \qquad \textbf{(1.36)}$$

The inverse transformation can be performed in a known manner and results in functions $E_y(x,s)$ and $E_z(x,s)$, which give the Fourier-spectrum of the field intensity at place x changing the notation s to $j\omega$. It seems from the Fourier-spectra that the solution at place x contains propagating and non-propagating terms. The inverse Fourier-transformation of the $s \to j\omega$ spectrum results in the space-time function of the solution. – Let us choose the $0 < \omega < \omega_b$ frequency-band in agreement with our original goal, i.e. we went to find an ELF-VLF, e.g. a whistler-mode solution in the following discussion, however, the method and the results are valid in this original form from DC to ∞, they have an extra wide band validity. – Applying the electric field intensities to the last two equations (1.29) we can find the components of the magnetic field intensity. (Let us mention that the solution is dimensionally also correct in each step of the derivation-process.)

After the $s \to j\omega$ transformation in the frequency band $0 < \omega < \omega_b$

$$p_1(\omega) = -jk_1(\omega),\ p_2(\omega) = jk_1(\omega),\ p_3 = -\alpha_3(\omega),\ p_4 = \alpha_3(\omega), \qquad \textbf{(1.37a)}$$

where

$$k_1(\omega) = \frac{1}{c}\sqrt{\frac{\omega\,\omega_b\omega_p^2 + \omega^2\left(\omega_p^2+\omega_b^2-\omega^2\right)}{\omega_b^2-\omega^2}} = k_0\sqrt{1+\frac{\omega_p^2}{\omega(\omega_b-\omega)}},$$

$$\alpha_3(\omega) = \frac{1}{c}\sqrt{\frac{\omega\,\omega_b\omega_p^2 - \omega^2\left(\omega_p^2+\omega_b^2-\omega^2\right)}{\omega_b^2-\omega^2}} = k_0\sqrt{\frac{\omega_p^2}{\omega(\omega_b+\omega)}-1}, \qquad \textbf{(1.37b)}$$

where $k_0 = \omega/c$ is the propagation factor is vacuum, $k_1(\omega)$ and $\alpha_3(\omega)$ are real quantities and during the discussion of roots we apply the known boundary-values of ω_b and ω_p in the plasmasphere of the Earth (see e.g. *Rishbeth and Garriott*, 1969); f_b ~ a few kHz to not more than 1 MHz, f_p ~ to cca. one hundred kHz to a few MHz and $f_b < f_p$. (In the case of other planets, the interplanetary space or artificial situations etc. the result of discussion depends on the actual plasma-parameters of the given case.) It is clear from relations (1.37b) that the first term of the solution propagates in the $+x$ direction, the second one propagates backward, in the $-x$ direction and the third and fourth terms are non-propagating terms in this most simple case. As an example for that let us see one of the components

$$E_y(x,\omega)=c^2\frac{a_\infty(\omega)a_0(\omega)a_2(\omega)}{2b_3(\omega)}\frac{1}{k_1^2(\omega)+\alpha_3^2(\omega)}\left\{\frac{\frac{a_1(\omega)}{a_2(\omega)}-jk_1(\omega)}{jk_1(\omega)}\cdot e^{-jk_1(\omega)x}-\right.$$

$$\left.-\frac{\frac{a_1(\omega)}{a_2(\omega)}+jk_1(\omega)}{jk_1(\omega)}\cdot e^{jk_1(\omega)x}-\frac{\frac{a_1(\omega)}{a_2(\omega)}-\alpha_3(\omega)}{\alpha_3(\omega)}\cdot e^{-\alpha_3(\omega)x}+\frac{\frac{a_1(\omega)}{a_2(\omega)}+\alpha_3(\omega)}{\alpha_3(\omega)}\cdot e^{\alpha_3(\omega)x}\right\}. \tag{1.38}$$

The form of the other components can be obtained in analogous manner and it can be seen easily, that in the whistler mode (in the forward propagating mode) the equation of the first term of (1.38), which propagates in the $+x$ direction, reads as

$$E_{yw}(x,\omega)=+jE_{zw}(x,\omega). \tag{1.39}$$

The magnetic components can be obtained from the second group of equations (1.29) by applying equations (1.38), where

$$H_x(p,s)=0,$$
$$H_y(p,s)=\frac{p}{\mu_0 s}E_z(p,s), \tag{1.40}$$
$$H_z(p,s)=-\frac{p}{\mu_0 s}E_y(p,s).$$

Therefore – as an example – the magnetic field belonging to the *first* term of the electric field in (1.38) is

$$H_{zw}(x,\omega)=-c^2\frac{a_\infty(\omega)a_0(\omega)}{2\mu_0 j\omega b_3(\omega)}\frac{[-jk_1(\omega)][a_1(\omega)-jk_1(\omega)a_2(\omega)]}{jk_1(\omega)}\cdot e^{-jk_1(\omega)x}. \tag{1.41}$$

Knowing the spectra $E_i(x,\omega)$ and $H_i(x,\omega)$ the signal can be determined. In doing so, it is to be taken into consideration, however – see the relations (1.37b) –, that the given modes exist in the band

$$0<\omega<\omega_b. \tag{1.42}$$

(In the same manner it is possible to discuss the propagating modes of the bands above ω_p and in the transition band $\omega_b<\omega<\omega_p$.) Finally, performing the inverse Fourier-transform, for instance the electrical components of the whistler-mode, propagating in the $+x$ direction will be given by

$$E_i(x,t)\cong\frac{1}{2\pi}\int_{-\infty}^{+\infty}E_{i0}(x,\omega)\cdot e^{j\varphi_0(\omega)}e^{j[\omega t-k_1(\omega)x]}d\omega, \tag{1.43}$$

which can be computed in the given case as

$$E_i(x,t)\cong\frac{1}{\pi}\int_{0}^{+\infty}E_{i0}(x,\omega)\cos[\omega t-k_1(\omega)x+\varphi_0(\omega)]d\omega, \tag{1.44}$$

where E_i is real, $\varphi_0(\omega)$ is the phase and $E_{i0}(x,\omega)$ is the absolute value of the complex spectral amplitude. In practical modelling we shall use $\omega_{\min}>0$ and $\omega_{\max}<+\infty$ boundary values (see Eq. (1.70) too), where $\omega_{\min}>0$ and $\omega_{\max}<\omega_b$ in the discussed frequency band.

The propagating signals and all existing modes in half-space "2" are described by equations (1.37) - (1.44), if we use the $x\to(x-x_0)$ substitutions, as a consequence of the position of the boundary surface A_i (see Fig. 1.6).

1.3.4 Derivation of the existing modes in the half-space "1

Let us suppose that the excitation current density generated by the lightning is flat and

$$\bar{J}_1(x,t)=J_0(x,t)\cdot\bar{e}_z. \tag{1.45}$$

Further the half-space "1" is infinite in the $-x$ direction and therefore reflected signals from this direction cannot exist, while at the $x=x_0$ surface reflected signals may exist. The task is to derive the solution of equation (1.10) in this case, i.e.

$$\overline{\nabla}\times\overline{H}_1=\overline{J}_1(x,t)+\varepsilon_0\frac{\partial\overline{E}_1}{\partial t},$$
$$\overline{\nabla}\times\overline{E}_1=-\mu_0\frac{\partial\overline{H}_1}{\partial t}. \tag{1.46}$$

The consequence of equation (1.45) and of the form of the whistler-mode solution of equations (1.30) is that within the possible solutions of equations (1.46) $\partial/\partial y=\partial/\partial z=0$.

The exciting signal in (1.46) generated by $\overline{J}_1$ is described by the following equations

$$\frac{\partial H_{1y}}{\partial x}=J_0(x,t)+\varepsilon_0\frac{\partial E_{1z}}{\partial t},$$
$$\frac{\partial E_{1z}}{\partial x}=\mu_0\frac{\partial H_{1y}}{\partial t}. \tag{1.47}$$

The other possible modes in "1" (e.g. reflected signals) are the solutions of the following homogeneous differential equations

$$\frac{\partial H_{1y}}{\partial x}=\varepsilon_0\frac{\partial E_{1z}}{\partial t},$$
$$\frac{\partial E_{1z}}{\partial x}=\mu_0\frac{\partial H_{1y}}{\partial t},$$

and (1.48)

$$\frac{\partial H_{1z}}{\partial x}=-\varepsilon_0\frac{\partial E_{1y}}{\partial t},$$
$$\frac{\partial E_{1y}}{\partial x}=-\mu_0\frac{\partial H_{1z}}{\partial t}.$$

The form of the possible solutions is arbitrary, therefore the methodology presented previously in points 1.3.3 - 1.3.4 can be applied successfully and the results are

$$H_{1y}(p,s)=\frac{p}{\mu_0 s}\cdot E_{1z}(p,s) \text{ and } E_{1z}(p,s)=\frac{\mu_0 s}{p^2-s^2/c^2}\cdot J_0(p,s). \tag{1.49}$$

Let $p=-\mathrm{j}k$ and $s=\mathrm{j}k$ be substituted and let be

$$k_0=\omega/c, \qquad Z_0=\sqrt{\mu_0/\varepsilon_0}=120\pi(\Omega) \qquad \text{and} \qquad c=1/\sqrt{\varepsilon_0\mu_0}.$$

Applying the previously defined values in equations (1.49) and deriving E_{1z}

$$E_{1z}(p,s)=\frac{Z_0}{2}\cdot\frac{J_0(p,s)}{p-\frac{s}{c}}-\frac{Z_0}{2}\cdot\frac{J_0(p,s)}{p+\frac{s}{c}}. \tag{1.50}$$

It is clear from equation (1.50) that J_0 generates two signals propagating in the $+x$ and $-x$ directions, respectively. We are interested only in the signal propagating in the $+x$ direction which is capable to excite the half-space "2". After the inverse-transformation the solution is

$$\begin{aligned} E_{1z}(x,t)&=-\frac{Z_0}{2}\int_0^{x_0} J_0\left(\xi;t-\frac{x-\xi}{c}\right)\mathrm{d}\xi,\\ H_{1y}(x,t)&=\frac{1}{2}\int_0^{x_0} J_0\left(\xi;t-\frac{x-\xi}{c}\right)\mathrm{d}\xi. \end{aligned} \tag{1.51}$$

and J_0 differs from zero only if $0\le t\le t_0$ and $0\le x\le x_0$ (see Appendix D).

The solutions of equation (1.48) are generally known. For example, from the first two equations of (1.48) we can get

$$\frac{\partial^2 E_{1z}}{\partial x^2}=\varepsilon_0\mu_0\frac{\partial^2 E_{1z}}{\partial t^2}=\frac{1}{c^2}\frac{\partial^2 E_{1z}}{\partial t^2}. \tag{1.52}$$

From (1.52)

$$\left(p^2-\frac{s^2}{c^2}\right)E_{1z}(p,s)=0 \qquad \text{and} \qquad p=\pm\frac{s}{c}, \tag{1.53}$$

and after the $s\leftrightarrow \mathrm{j}\omega$ transformation

$$p_{1,2} = \pm \mathrm{j}\frac{\omega}{c} \triangleq \mathrm{j}k_{1,2},$$

and from this relation

$$k_{1,2} = \pm\frac{\omega}{c} \equiv \pm k_0. \tag{1.54}$$

Using (1.54) we can get the $E_{1z}(x,t)$ function as

$$E_{1z}(x,t) = \frac{1}{2\pi}\int_{-\infty}^{+\infty} E_{1z}(\omega)\mathrm{e}^{\mathrm{j}\omega\left(t \mp \frac{x}{c}\right)}\mathrm{d}\omega . \tag{1.55}$$

We can get the $H_{1z}(x,t)$ on the same way from (1.49) and the $E_{1y}(x,t)$ and $H_{1z}(x,t)$ functions on an analogous manner.

1.3.5 Mode-coupling at the boundary surface A_l

To find the final form of the propagating signal we must use equations (1.6), (1.8) and (1.11) in the given case. The actual form of equation (1.6) is

$$s_1(\bar{r}) = 1 - \Theta(x - x_0) \quad \text{and} \quad s_2(\bar{r}) = \Theta(x - x_0), \tag{1.56}$$

and therefore from (1.8)

$$\overline{\nabla} s_1(\bar{r}) = -\delta(x - x_0)\cdot \bar{e}_x \quad \text{and} \quad \overline{\nabla} s_2(\bar{r}) = \delta(x - x_0)\cdot \bar{e}_x . \tag{1.57}$$

Deriving equations (1.11) in the given case one obtains that at the place $x = x_0$

$$\begin{aligned}
&\sum_{i=1}^{n_1} H_{1zi}(x_0,t) - \sum_{j=1}^{n_2} H_{2zj}(x_0,t) = 0,\\
&-\sum_{i=1}^{n_1} H_{1yi}(x_0,t) + \sum_{j=1}^{n_2} H_{2yj}(x_0,t) = 0, \qquad (1.58)\\
&\sum_{i=1}^{n_1} E_{1zi}(x_0,t) - \sum_{j=1}^{n_2} E_{2zj}(x_0,t) = 0, \qquad \textit{cont.}
\end{aligned}$$

$$-\sum_{i=1}^{n_1} E_{1yi}(x_0,t)+\sum_{j=1}^{n_2} E_{2yj}(x_0,t)=0,$$

$$-\varepsilon_0\sum_{i=1}^{n_1} E_{1xi}(x_0,t)+\sum_{j=1}^{n_2} D_{2xj}(x_0,t)=0, \qquad \textbf{(1.58)}$$

$$-\sum_{i=1}^{n_1} H_{1xi}(x_0,t)+\sum_{j=1}^{n_2} H_{2xj}(x_0,t)=0,$$

In the last two equations of (1.58) all components are definitely zero – see (1.29) and (1.30a) – and the first four equations remain only. Solving these four equations of (1.58) it is possible to derive the signal (the whistler-mode) propagating in the direction in medium "2". One may derive the functions $E_{2zw0}(x=x_0,\omega)$ and $H_{2yw0}(x=x_0,\omega)$ at $x=x_0$ from the Laplace-transformed equations and is known from the earlier that

$$E_{2yw}(x,t)=+\mathrm{j}E_{2zw}(x,t); \qquad H_{2zw}(x,t)=-\mathrm{j}H_{2yw}(x,t). \qquad \textbf{(1.59)}$$

In the following the w index means the propagating signals (whistler-mode) and the N index means the second mode of the signal, which is non propagating in this case. We must apply in equation-system (1.58) all of the existing modes in half-spaces "1" and "2". As we derived in the previous points, these modes are in half-space "1" the mode E_{1zG} generated directly by source-current $\overline{J}_{si}$ – see (1.51) – and the possible reflected signals E_{1yR} and E_{1zR} and their magnetic pairs – see (1.55). Therefore

$$E_{1zG}=-\frac{Z_0}{2}\int_0^{x_0} J_0\left(\xi,t-\frac{x-\xi}{c}\right)\mathrm{d}\xi \quad \text{and} \quad H_{1yG}=-\frac{1}{Z_0}E_{1zG}\,. \qquad \textbf{(1.60)}$$

The reflected signals are

$$E_{1yR}=\frac{1}{2\pi}\int_{-\infty}^{+\infty} E_{1yR}(\omega)\cdot \mathrm{e}^{\mathrm{j}\omega\left(t+\frac{x}{c}\right)}\mathrm{d}\omega \quad \text{and} \quad H_{1zR}=-\frac{1}{Z_0}E_{1yR},$$

$$E_{1zR}=\frac{1}{2\pi}\int_{-\infty}^{+\infty} E_{1zR}(\omega)\cdot \mathrm{e}^{\mathrm{j}\omega\left(t+\frac{x}{c}\right)}\mathrm{d}\omega \quad \text{and} \quad H_{1yR}=\frac{1}{Z_0}E_{1zR}. \qquad \textbf{(1.61)}$$

In the half-space "2" we found one propagating and one non-propagating, i.e. one whistler (E_{2zw}) and one non-whistler (E_{2zN}) mode – see (1.38) and (1.43) together

$$
\begin{aligned}
E_{2zw} &= \frac{1}{8\pi}\int_{-\infty}^{+\infty}\left[\mathrm{j}\frac{e'_{2z0t}(\omega)}{k_1(\omega)}+e_{2z0t}(\omega)\right]\mathrm{e}^{\mathrm{j}[\omega t-k_1(\omega)(x-x_0)]}\mathrm{d}x,\\
E_{2zN} &= -\frac{1}{8\pi}\int_{-\infty}^{+\infty}\left[\frac{e'_{2z0t}(\omega)}{\alpha_3(\omega)}-e_{2z0t}(\omega)\right]e^{\mathrm{j}\omega t-\alpha_3(\omega)(x-x_0)}\mathrm{d}x,\\
E_{2yw} &= \mathrm{j}E_{2zw},\\
E_{2yN} &= -\mathrm{j}E_{2zN},\\
H_{2yw} &= -\frac{1}{8\pi Z_0}\int_{-\infty}^{+\infty}\left[\mathrm{j}\frac{e'_{2z0t}(\omega)}{k_0(\omega)}+\frac{k_1(\omega)}{k_0(\omega)}e_{2z0t}(\omega)\right]\mathrm{e}^{\mathrm{j}[\omega t-k_1(\omega)(x-x_0)]}\mathrm{d}x,\\
H_{2yN} &= -\frac{1}{8\pi Z_0}\int_{-\infty}^{+\infty}\left[\frac{e'_{2z0t}(\omega)}{k_0(\omega)}-\frac{\alpha_3(\omega)}{k_0(\omega)}e_{2z0t}(\omega)\right]\mathrm{e}^{\mathrm{j}\omega t-\alpha_3(\omega)(x-x_0)}\mathrm{d}x,\\
H_{2zw} &= -\mathrm{j}H_{2yw},\\
H_{2zN} &= \mathrm{j}H_{2yN}.
\end{aligned}
\tag{1.62}
$$

Applying relations (1.60), (1.61) and (1.62) in the remained four equations of (1.58)

$$
\begin{aligned}
&-\frac{1}{Z_0}E_{1yR}(x_0,t)=-\mathrm{j}H_{2yw}(x_0,t)+\mathrm{j}H_{2yN}(x_0,t),\\
&-\frac{1}{Z_0}E_{1zG}(x_0,t)+\frac{1}{Z_0}E_{1zR}(x_0,t)=H_{2yw}(x_0,t)+H_{2yN}(x_0,t),\\
&E_{1zG}(x_0,t)+E_{1zR}(x_0,t)=E_{2zw}(x_0,t)+E_{2zN}(x_0,t),\\
&E_{1yR}(x_0,t)=\mathrm{j}E_{2zw}(x_0,t)-\mathrm{j}E_{2zN}(x_0,t).
\end{aligned}
\tag{1.63}
$$

After solving equation-system (1.63) we get a relation between the propagating and non-propagating signals in "2"

$$
\left[1+\frac{k_1(\omega)}{k_0(\omega)}\right]\left[\frac{e'_{2z0t}(\omega)}{k_1(\omega)}-\mathrm{j}e_{2z0t}(\omega)\right]=\left[1-\mathrm{j}\frac{\alpha_3(\omega)}{k_0(\omega)}\right]\left[\mathrm{j}\frac{e'_{2z0t}(\omega)}{\alpha_3(\omega)}-\mathrm{j}e_{2z0t}(\omega)\right].\tag{1.64}
$$

Applying (1.64) in (1.63) and deriving the sum of the first and fourth equations – which is identical to (1.64) – and deriving the difference of the third and second equations we get

$$2E_{1zG}(x_0,t)=E_{2zw}(x_0,t)-Z_0\cdot H_{2yw}(x_0,t)+E_{2zN}(x_0,t)-Z_0\cdot H_{2yN}(x_0,t),$$
$$E_{2zw}(x_0,t)-Z_0\cdot H_{2yw}(x_0,t)=E_{2zN}(x_0,t)-Z_0\cdot H_{2yN}(x_0,t).$$

Therefore

$$E_{1zG}(x_0,t)=E_{2zw}(x_0,t)-Z_0\cdot H_{2yw}(x_0,t). \tag{1.65}$$

Applying the formulas of (1.62) in equation (1.65) and using the notation

$$A(\omega)=\frac{e'_{2z0t}}{k_1(\omega)}-\mathrm{j}e_{2z0t}(\omega),$$

we get

$$-\frac{Z_0}{2}\int_0^{x_0}J_0\left(\xi,t-\frac{x_0-\xi}{c}\right)\cdot\mathrm{d}\xi=\frac{\mathrm{j}}{8\pi}\int_{-\infty}^{+\infty}A(\omega)\mathrm{e}^{\mathrm{j}\omega t}\mathrm{d}\omega+\frac{\mathrm{j}}{8\pi}\int_{-\infty}^{+\infty}\frac{k_1(\omega)}{k_0(\omega)}A(\omega)\mathrm{e}^{\mathrm{j}\omega t}\mathrm{d}\omega,$$

i.e.

$$\mathrm{j}2Z_0\int_0^{x_0}J_0\left(\xi,t-\frac{x_0-\xi}{c}\right)\cdot\mathrm{d}\xi=\frac{1}{2\pi}\int_{-\infty}^{+\infty}\frac{k_0(\omega)+k_1(\omega)}{k_0(\omega)}A(\omega)\mathrm{e}^{\mathrm{j}\omega t}\mathrm{d}\omega. \tag{1.66}$$

Let us define the Fourier-transform of the left side of equation (1.66) as

$$I_{x0}(\omega)\triangleq\int_{-\infty}^{+\infty}\left\{\int_0^{x_0}J_0\left(\xi,t-\frac{x_0-\xi}{c}\right)\cdot\mathrm{d}\xi\right\}\mathrm{e}^{-\mathrm{j}\omega t}\cdot\mathrm{d}t. \tag{1.67}$$

(The dimension of $I_{x0}(\omega)$ is As/m, as we know.) Applying $I_{x0}(\omega)$ in equation (1.66)

$$A(\omega)=\mathrm{j}2Z_0\frac{k_0(\omega)}{k_0(\omega)+k_1(\omega)}I_{x0}(\omega), \tag{1.68}$$

and therefore

$$E_{2zw}(x,t) = -\frac{Z_0}{4\pi}\int_{-\infty}^{+\infty} I_{x0}(\omega)\frac{k_0(\omega)}{k_0(\omega)+k_1(\omega)} e^{j[\omega t - k_1(\omega)(x-x_0)]}d\omega,$$
$$H_{2yw}(x,t) = \frac{1}{4\pi}\int_{-\infty}^{+\infty} I_{x0}(\omega)\frac{k_1(\omega)}{k_0(\omega)+k_1(\omega)} e^{j[\omega t - k_1(\omega)(x-x_0)]}d\omega. \quad \textbf{(1.69)}$$

The real space-time functions of the solution are given in the form of (1.44), i.e.

$$E_{2zw}(x,t) \cong -\frac{Z_0}{2\pi}\int_{\omega_{min}}^{\omega_{max}} \frac{k_0(\omega)}{k_0(\omega)+k_1(\omega)}\cdot A_I(\omega)\cdot \cos[\omega t - k_1(\omega)(x-x_0) + \varphi_I(\omega)]\cdot d\omega \quad \textbf{(1.70)}$$

etc. In equation (1.70)

$A_I(\omega) \equiv \text{Amplitude}\,[I_{x0}(\omega)]$ and
$\varphi_I(\omega) \equiv \text{Phase}\,[I_{x0}(\omega)]$.

From theoretical point of view the integration boundaries in equation (1.70) are $0 \le \omega \le +\infty$. However, from practical point of view we must exclude the place of singularities and in the case of an ELF~VLF application it is enough to integrate only through these frequency bands. In the given case $\omega = 0$ and $\omega = \omega_b < \omega_p$ are the places of singularities. Therefore in (1.70) $\omega_{min} > 0$ and $\omega_{max} < \omega_b$. In some modelling task ω_{min} and ω_{max} must be identical the boundary frequencies of the measuring system to compare the calculated and measured signals.

1.3.6 Character of excitations

In the relation (1.67) the form of excitation is arbitrary. However, one excitation is extremely important from theoretical point of view and a few others are important for practical modelling.

a) The Dirac-delta excitation:

This excitation has a special importance because the answer – in this case (1.69) or (1.70) – given by a system, which is excited by a Dirac-delta distribution, is the *transfer function*. The transfer function (see e.g. *Simonyi*, 1963) characterises the nature of the excited system. In our case this means that the transfer function (Dirac-answer or transfer function response) characterises the propagation itself, and therefore it characterises the medium traversed by the signal. This open an exact way to discriminate between the effects of propagation ("propagation path") itself and of the other conditions (the structure of excitation, signal interference etc.). In this case

$$J_0(x,t) = I_0 \cdot \delta(x) \cdot \delta(t), \tag{1.71a}$$

and therefore

$$I_{x0} = I_0 = \text{constant}. \tag{1.71b}$$

Determining the transfer function we apply the $I_0 = 1$ (As/m) value.

b) Other useful excitations:

The most simple rectangular-impulse excitation is a rectangular current density appearing at the place $x = x_0$ in the medium "1". The current density is

$$J_0(x,t) = I_0[\Theta(t) - \Theta(t - t_0)] \cdot \delta(x - x_0), \tag{1.72a}$$

where $I_0 = \text{constant}$ (A/m). After deriving the equation (1.67) and knowing that the spectral density value at the place $\omega = 0$ has no role in the equation (1.70) – it would be only a non-propagating DC component, and therefore we can exclude it from the further investigation – the excitation is characterised by

$$I_{x0}(\omega) = \frac{2I_0}{\omega}\sin\left(\frac{\omega t_0}{2}\right) \cdot \mathrm{e}^{-\mathrm{j}\omega\frac{t_0}{2}}. \tag{1.72b}$$

However, it is possible that the source place of the rectangular-impulse excitation is not at the boundary surface but at the place $x = x_*$, i.e.

$$J_0(x,t) = I_0[\Theta(t) - \Theta(t - t_0)] \cdot \delta(x - x_0), \tag{1.73a}$$

where $x_* \leq x_0$ must fulfil. In this case, the only variation in $I_{x0}(\omega)$ is that a time delay appears in (1.72b), i.e.

$$I_{x0}(\omega) = \frac{2I_0}{\omega}\sin\left(\frac{\omega t_0}{2}\right)\mathrm{e}^{-\mathrm{j}\omega\left(\frac{t_0}{2} + \frac{x_0 - x_*}{c}\right)}. \tag{1.73b}$$

In some practical cases, a more realistic modelling of the impulse-like excitation is necessary. In these cases, it is possible to use a more realistic square impulse excitation in space and time, i.e.

$$J_0(x,t) = I_0[\Theta(x) - \Theta(x - x_n)][\Theta(t) - \Theta(t - t_0)], \tag{1.74a}$$

where x_n is the length of impulse in space and $0 \le x_n \le x_0$ must fulfil, and t_0 is the length of impulse in time, and the dimension of I_0 in this case is (A/m²). After deriving the equation (1.67) this excitation is characterised by

$$I_{x0}(\omega) = I_0 \frac{c}{\omega}\left\{\frac{1}{\omega}\sum_{i=1}^{n}(-1)^i e^{-j\omega t_i} + e^{j\frac{\pi}{2}}\left[\left(t_3 - t_4 + \frac{x_t}{c}\right)e^{-j\omega t_4} + \left(t_2 - t_1 - \frac{x_t}{c}\right)e^{-j\omega t_2}\right]\right\}, \quad \textbf{(1.74b)}$$

where

$$x_t = \begin{cases} ct_0, & \text{if } t_0 \le \dfrac{x_n}{c}; \\ x_n, & \text{if } t_0 \ge \dfrac{x_n}{c}; \end{cases}$$

and

$$t_1 = \frac{x_0 - x_n}{c}; \qquad t_2 = \begin{cases} \dfrac{x_0 - x_n}{c} + t_0, & \text{if } t_0 \le \dfrac{x_n}{c}; \\ \dfrac{x_0}{c}, & \text{if } t_0 \ge \dfrac{x_n}{c}; \end{cases}$$

$$t_3 = \begin{cases} \dfrac{x_0}{c}, & \text{if } t_0 \le \dfrac{x_n}{c}; \\ \dfrac{x_0 - x_n}{c} + t_0, & \text{if } t_0 \ge \dfrac{x_n}{c}; \end{cases} \qquad t_4 = \frac{x_0}{c} + t_0 .$$

Besides this excitation, which we can use easily in practical model investigations, it is possible to apply the measured spectrums of lightning, i.e. spherics (e.g. *Tixier*, 1976). At the modelling of active satellite measurements we can use the exact time-function of excitation applied onboard of a given satellite. It is clear that the applicability of these results is flexible and it is succesful on a wide range of propagation phenomena.

1.3.7 Some results of model-calculations

In a model-calculation the first result is the space-time function of the propagating signal, or more precisely it is the time function of the propagating signal in a given place, at a given distance from the origo or from the boundary surface A_1. After this step we can use the known analytical methods of whistlers or other signals, which are used during the analysis of real signals measured on Earth or onboard of satellites or in other cases. (See *Bognár et al.*, 1995). These analytical methods are presented in Chapter 4. However, we shall present results here obtained by applying these methods at the final phase of analysis of every interesting problem-solution.

In this case, the following spectra were used as excitation: the Dirac-delta excitation, the rectangular square impulse and real spectra of spherics. The measured waveforms of spherics given by *Tixier* (1976) have in most cases a duration between 0.1 and 2.0 ms. For our studies, we selected three measured waveforms, which will be referred to as lightning 1, 2 and 3. The length of the rectangular impulse (in time) is t_0, namely the $t_0 = 0.25$ ms was considered as "short impulse" and $t_0 = 2$ ms as "long impulse"; further $x_* = ct_0$ was chosen as the length of the impulse in the space which was a non-characteristic parameter.

The time-function of the whistler-signal at the place $x \cong 30.000$ km in the magnetoionic medium ($\omega_b \cong 55 \cdot 10^3$ 1/s, i.e. $f_b \cong 8.75$ kHz and $\omega_p \cong 0{,}734 \cdot 10^6$ 1/s, i.e. $f_p \cong 0.12$ MHz) was determined to compare the calculated results to the known results of the analysis of measured whistlers (e.g. *Tarcsai*, 1981; *Hamar* et al., 1992).

The forms of the calculated signals at the place x are shown in Figs. 1.7 and 1.8. Fig. 1.7 shows the whistler-signals of the following excitations: a Dirac-delta, a short and a long rectangular square-impulse, respectively. Fig. 1.8 shows the whistler signals arising from the following excitations: three different spectra of lightning (spherics) and a half-sinusoidal current density – computing directly by equation (1.67). This figures present the details of the front and the first part of the calculated whistlers too.

Fig. 1.9 shows the FFT-spectrum of a calculated whistler of the "short impulse" excitation at the places $x/2$, x and $2x$ demonstrating the increasing dispersion. The FFT pattern of calculated whistlers in this homogeneous, simple magnetoionic model is in an acceptable agreement with the known, measured whistler-spectra, also in the vicinity of the nose-frequency, as was expected based upon the theory and Figs. 1.7 and 1.8 too. (The nose appears at a frequency of $f_b/4$, as was expected. The nose frequencies for various excitations derived from the computed or measured space-time functions are the following: 2257±1 Hz for the Dirac-delta; 2256±1 Hz for "long impulse"; 2250±1 Hz for "short impulse"; 2260±1 Hz for lightning 1; 2283±5 Hz for whistler No 89221 EI, measured at Halley station.) The general agreement between the measured signals and the computed ones is evident now, if we use only a simple homogeneous medium-model.

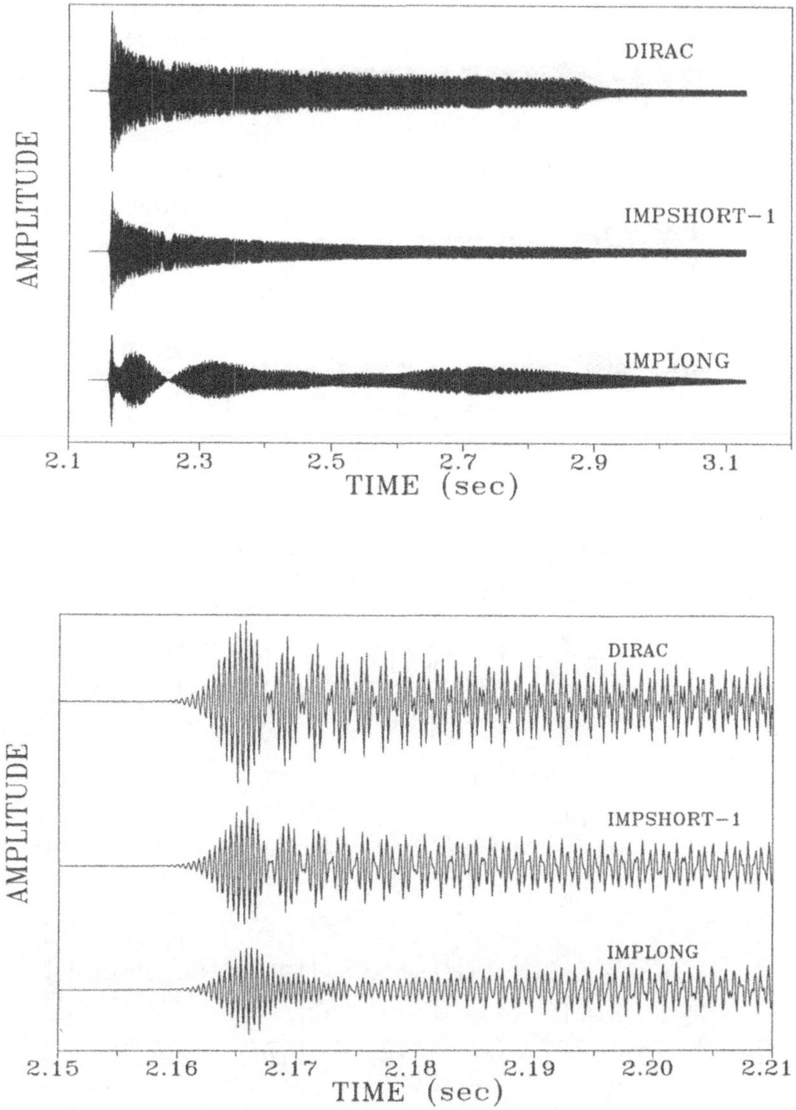

Figure 1.7. The time-functions of calculated whistlers at the place *x* of different excitations (the full computed functions and the front parts).

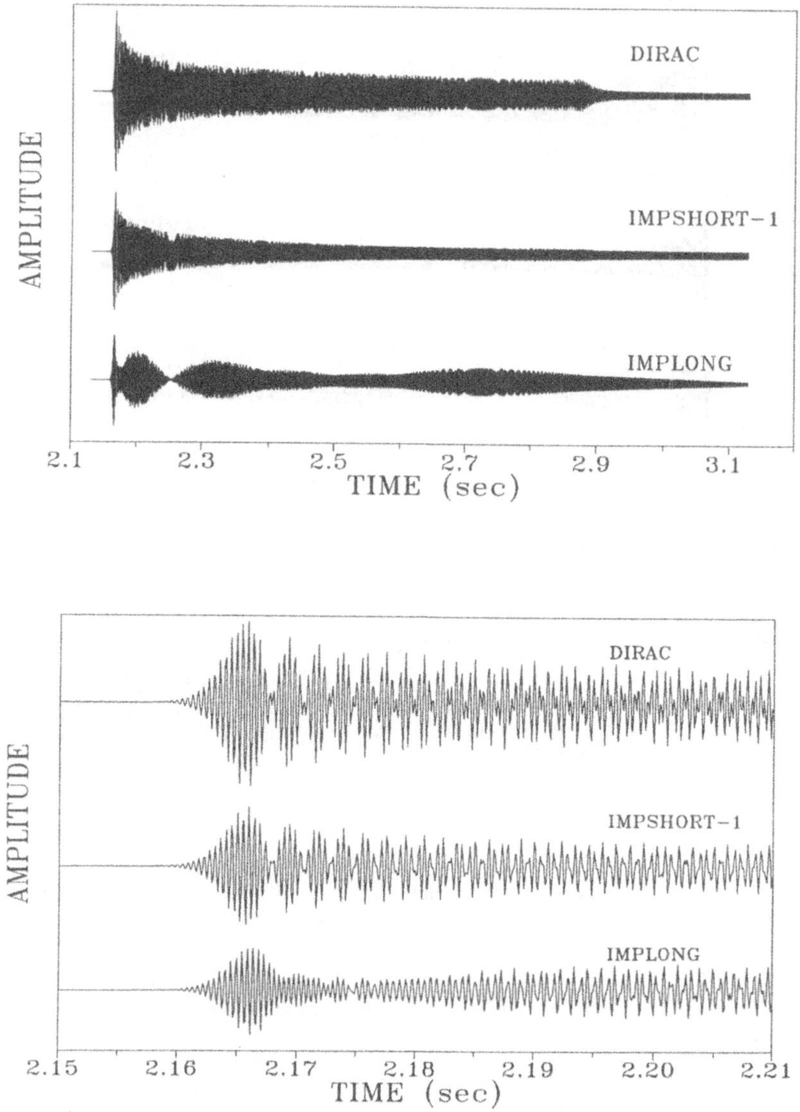

Figure 1.8. The time-functions of calculated whistlers at the place x of different excitations (the full computed functions and the front parts).

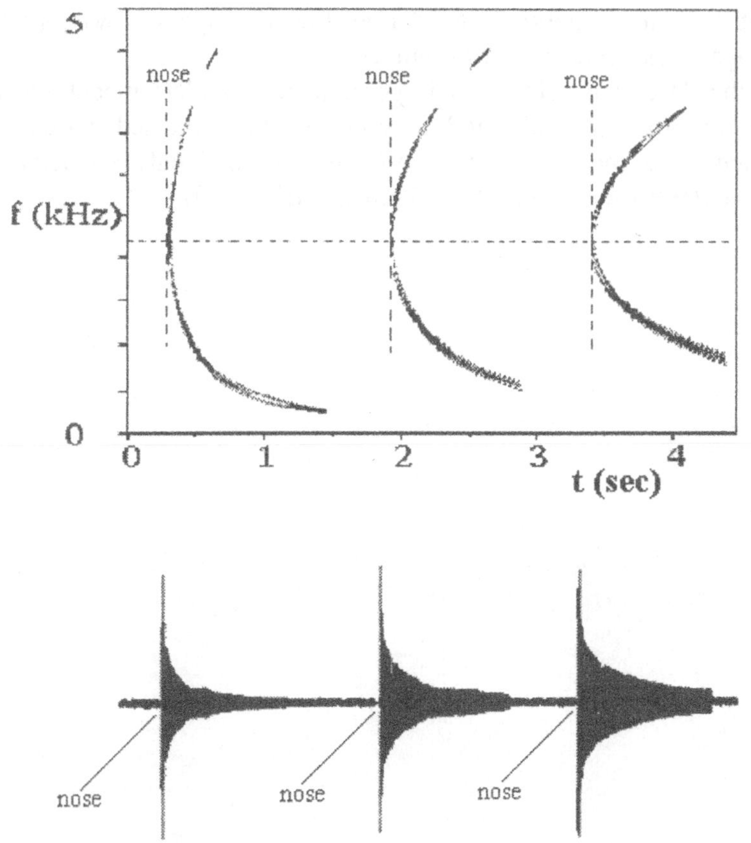

Figure 1.9. The calculated whistler signal excited by a "short impulse" on three different places $(x_1 = x/2;\ x_2 = x;\ x_3 = 2x)$

1.3.8 Conclusions

As was seen above, the method – presented in point 1.1 and applied in point 1.3 – is suitable to find the solution of Maxwell's equations in the cases of non-monochromatic, general shape signals or excitations. This was tested successfully to find the general whistler-mode solution in a homogeneous, anisotropic, cold, magnetoionic medium in the case of strictly longitudinal propagation.

However, the method used in solving Maxwell's equations is more general and therefore it is usable in quasi-longitudinal or arbitrarily directed propagation or in cases in which $\overline{B}_{F0}$ is non-perpendicular to the boundary plain or in which the boundary surface is not flat, or any other cases too.

Also, the closed forms of the solution – see formulas (1.69) and (1.70) – suggest that it is possible to derive the propagating wave-patterns in inhomogeneous media or in waveguides.
Using the described solution it is possible to calculate model-whistlers, to analyse the fine structure and to compare the obtained results of such investigations to the results obtained for measured signals (e.g. *Hamar* et al., 1990; *Lichtenberger* et al., 1991; *Ferencz et al.,* 1996)

Chapter 2

APPLICATION OF THE METHOD FOR DIFFERENT PROPAGATIONAL SITUATIONS

2.1 Propagation in inhomogeneous, lossless, electron-plasma – longitudinal propagation

The method presented in Chapter 1 can be used not only in cases in which the media in space-segments are homogeneous between the boundary surfaces – see Fig. 1.1 –, but also in cases in which the media in these space-segments are weakly inhomogeneous (quasi-homogeneous) or inhomogeneous (e.g. *Ferencz,* 1995b) and the gradient of the inhomogeneous medium parameters is limited only by the validity-condition of the method (*Ferencz,* 1978a). In the following let us generalise the method of solution – presented in point 1.1 and applied in point 1.3 – for the cases in which the media in the space-segments are quasi-homogeneous and inhomogeneous between the boundary surfaces.

The example will be presented in a case which has a geometry presented in Fig. 1.6 and in which the medium in volume "2" is inhomogeneous. The medium in volume "1" and the place of the boundary plane A_1 remains unchanged. Therefore the A_1 plane at $x = x_0$ is perpendicular to the axis x, and $\overline{n}_0 = \overline{e}_x$. Let the medium be in half-space (volume) "2" an inhomogeneous magnetoionic medium.

2.1.1 The propagational problem in half-space "2"

The static magnetic induction $\overline{B}_{F0} = B_{F0} \cdot \overline{e}_x$ is directed parallel to the axis $+x$. This is a commonly used, but strong simplification, because the inhomogeneous magnetic field must change its direction too. However, the models using this simplification are practically usable. As in the earlier investigation the plasma is linear, tempered, cold, magnetised, lossless, time-invariant, single component electron-plasma with electron density N, m

is the mass and q is the charge of an electron (e is the absolute value of the charge of an electron, $q=-e$) and $\bar{v}$ is the average velocity of electrons excited by the electromagnetic field, respectively. The inhomogeneity is caused by the change in B_{F0} and N, written as

$$B_{F0} = B_{F0}(x) \quad \text{and} \quad N = N(x). \tag{2.1}$$

As a first approximation, the change of the direction of $\bar{B}_{F0}(x)$ is omitted. Let the solution be a plane wave, propagating exactly longitudinally, i.e. exactly along the direction $\bar{e}_x$ in the plasma. (The procedure can be generalised of course with some mathematical complications. We can use this magnetic field model in inhomogeneous case, because we shall investigate a strictly longitudinal propagation. See more in point 2.5.) Thus the well-known and widely accepted longitudinal propagation model is used to find a real full wave solution. Those effects which maintain the field aligned propagation and those which play only a secondary role in the formation of the space-time function of electromagnetic field strength components will not be treated here.

As a consequence of (2.1), unlike the homogeneous case, now both the angular electron gyrofrequency (ω_b) and the angular electron plasma frequency (ω_p) depend on the place, i.e.

$$\omega_b(x) = \frac{e}{m} B_{F0}(x) \quad \text{and} \quad \omega_p^2(x) = \frac{e^2}{\varepsilon_0 m} N(x). \tag{2.2}$$

The equations describing the interaction between the signal and the medium are formally identical with the previous ones (equations (1.25)~(1.26b)), that is

$$m\frac{\partial \bar{v}}{\partial t} = q\left(\bar{E} + \bar{v} \times \bar{B}_{F0}\right),$$

however, the coefficients in the component equations depend on space-coordinates. Rearranging (1.25), as usual

$$\frac{\partial v_x(\bar{r},t)}{\partial t} = \frac{q}{m} E_x(\bar{r},t),$$
$$\frac{\partial v_y(\bar{r},t)}{\partial t} + \omega_b(\bar{r}) \cdot v_z(\bar{r},t) = \frac{q}{m} E_y(\bar{r},t), \tag{2.3}$$
$$\frac{\partial v_z(\bar{r},t)}{\partial t} - \omega_b(\bar{r}) \cdot v_y(\bar{r},t) = \frac{q}{m} E_z(\bar{r},t).$$

The solution of these equations exists and is known in closed form (see e.g. *Kamke,* 1956, 1965), which is formally identical with that of the homogeneous case, as a consequence of the fortunate structure of equations (2.3) and of the time invariance of the medium, however ω_b and ω_p now *depend on place.* It follows that at the beginning the derivation steps of the homogeneous case may be applied using $\omega_b(x)$ and $\omega_p(x)$, instead of ω_b and ω_p in equations (1.27)~(1.30). Therefore

$$E_x = 0, \quad H_x = 0 \tag{2.4}$$

are valid in this case too, and the task is to find the components (E_y, E_z) and (H_y, H_z).

2.1.2 Longitudinal quasi-homogeneous plane wave solution in half-space "2"

The form of equation system (1.30b) is now

$$c^2 \frac{\partial^2 E_y}{\partial x^2} = \omega_p^2(x) \int_0^t \frac{\partial E_y}{\partial \tau} \cos[\omega_b(x)(t-\tau)] \mathrm{d}\tau -$$
$$- \omega_b(x) \cdot \omega_p^2(x) \int_0^t E_z \cos[\omega_b(x)(t-\tau)] \mathrm{d}\tau + \frac{\partial^2 E_y}{\partial t^2},$$
$$c^2 \frac{\partial^2 E_z}{\partial x^2} = \omega_p^2(x) \int_0^t \frac{\partial E_z}{\partial \tau} \cos[\omega_b(x)(t-\tau)] \mathrm{d}\tau + \tag{2.5}$$
$$+ \omega_b(x) \cdot \omega_p^2(x) \int_0^t E_y \cos[\omega_b(x)(t-\tau)] \mathrm{d}\tau + \frac{\partial^2 E_z}{\partial t^2}.$$

The shape of $\bar{E}$ may be general, the only known fact is, that it does not exist for $t < 0$, and for $x \geq x_0$ the propagating signal is to be found, so the idea of applying the Laplace-transform arises. as earlier

$$
\begin{aligned}
&x \longleftrightarrow p, \quad t \longleftrightarrow s,\\
&E_i(x,t=0) = e_{ix0}(x) \longleftrightarrow e_{ix0}(p),\\
&E_i(x=0,t) = e_{ix0}(t) \longleftrightarrow e_{i0t}(s),\\
&\left.\frac{\partial E_i(x,t)}{\partial x}\right|_{x=0} = e'_{i0t}(t) \longleftrightarrow e'_{i0t}(s),\\
&\left.\frac{\partial E_i(x,t)}{\partial t}\right|_{t=0} = e'_{ix0}(x) \longleftrightarrow e'_{ix0}(p),
\end{aligned}
\tag{2.6}
$$

where $i = y,z$. Examining the equation system (2.5), it can be pointed out, that the Laplace-transform according to t can be executed in the same manner as in the homogeneous case, however the one of $x \to p$ cannot be executed and is not even practical, as the form of $\omega_b(x)$ and $\omega_p(x)$ could be arbitrary, and furthermore, as a consequence of the structure of the equations after the $t \to s$ transform. Let

$$
E_i(x,s) = \int_0^\infty E_i(x,t)\cdot \mathrm{e}^{-st}\mathrm{d}t \tag{2.7}
$$

be introduced, then the Laplace-transform of equations (2.5) according to t is the following

$$
\begin{aligned}
c^2\frac{\partial^2 E_y(x,s)}{\partial x^2} &= s^2\left[\frac{\omega_p^2(x)}{s^2+\omega_b^2(x)}+1\right]E_y(x,s) -\\
&\quad -\omega_b(x)s\frac{\omega_p^2(x)}{s^2+\omega_b^2(s)}E_z(x,s),\\
c^2\frac{\partial^2 E_z(x,s)}{\partial x^2} &= s^2\left[\frac{\omega_p^2(x)}{s^2+\omega_b^2(x)}+1\right]E_z(x,s) +\\
&\quad +\omega_b(x)s\frac{\omega_p^2(x)}{s^2+\omega_b^2(s)}E_y(x,s) - e'_{zx0}(x),
\end{aligned}
\tag{2.8}
$$

where it is assumed, in agreement with the homogeneous case, that $e_{yx0}(x) \equiv e_{y0t}(t) \equiv e'_{yx0}(x) \equiv e'_{y0t}(t) \equiv e_{zx0}(x) \equiv 0$. Further, it follows from Maxwell's equations, that

$$
H_y(x,s) = \frac{1}{\mu_0 s}\frac{\partial E_z(x,s)}{\partial x} \quad \text{and} \quad H_z(x,s) = -\frac{1}{\mu_0 s}\frac{\partial E_y(x,s)}{\partial x}. \tag{2.9}
$$

Now let us use one of the methods which proved to be effective in the study of quasi-homogeneous, weakly inhomogeneous media – the *eikonal-equation*, the *W.K.B. solution* or the method given by *Ferencz* (1977a) – keeping in mind, that the signal to be found now is not a monochromatic one. The result will be the same, anyway. Let now the W.K.B. solution be applied.

Apart from the spatial dependence of ω_p and ω_b in equations (2.8) the known result can be obtained (point 1.3). Let therefore the solution of equations (2.8) be found for the case of weak inhomogeneities (*Budden*, 1966; *Ferencz*, 1977a) by generalising the homogeneous solution. That part of the wave propagating toward the $+x$ direction will be considered, in which we are particularly interested. It will be assumed that the magnetoionic, weakly inhomogeneous plasma fills only the half-space "2".

It is known (equations (1.35)~(1.38)), that the wave propagating along the $+x$-axis is attached to the value

$$p_1(x,s) = -\frac{1}{c}\sqrt{\frac{j\omega_b(x)\cdot\omega_p^2(x)s + s^2\left[\omega_p^2(x)+\omega_b^2(x)+s^2\right]}{\omega_b^2(x)+s^2}}, \qquad \textbf{(2.10)}$$

which is kept in unaltered form as a characteristic pole of the space Laplace-transform existing in the homogeneous case. According to the W.K.B. philosophy (Appendix E)

$$E_{2zw}(x,s) = \frac{a_2(s)p_1(x,s)+a_1(s)}{4p_1(x,s)}\cdot e^{\int_{x_0}^{x} p_1(\xi,s)d\xi}, \qquad \textbf{(2.11)}$$

where $a_1(s)$ and $a_2(s)$ are boundary conditions defined by the excitation functions as was seen in point 1.3 in equation (1.35)

$$a_1 = E_0^+ e'_{z0t}(s) \quad \text{and} \quad a_2 = E_0^+ e_{z0t}(s). \qquad \textbf{(2.12)}$$

Then

$$H_{2yw}(x,s) = \frac{p_1(x,s)}{\mu_0 s} E_{2zw}(x,s). \qquad \textbf{(2.13)}$$

Let now the $s \to j\omega$ conversion be executed and a generalised W.K.B. criterion be applied, which provides the magnitude of the spectral Poynting vector $S(x,\omega)$, which should be independent of place in this case, while the signal is non-monochromatic and the propagation is strictly linear. Here $\overline{S}(x,\omega)=\overline{E}(x,\omega)\times\overline{H}(x,\omega)$ and $S(x,\omega)$ is the magnitude of $\overline{S}(x,\omega)$.

So let the solution be

$$E_{2zw}(x,\omega)=E_{z0}^{+}\cdot \mathrm{e}^{-\mathrm{j}\int_{x0}^{x}k_1(\xi,\omega)\mathrm{d}\xi},$$
$$H_{2yw}(x,\omega)= \tag{2.14}$$
$$=-\frac{E_{z0}^{+}}{Z_0}\sqrt{\frac{\omega_b(x)\omega_p^2(x)\omega+\omega^2\left[\omega_p^2(x)+\omega_b^2(x)-\omega^2\right]}{\omega^2\left[\omega_b^2(x)-\omega^2\right]}}\cdot \mathrm{e}^{-\mathrm{j}\int_{x0}^{x}k_1(\xi,\omega)\mathrm{d}\xi},$$

and

$$k_1(x,\omega)=\frac{1}{c}\sqrt{\frac{\omega_b(x)\omega_p^2(x)\omega+\omega^2\left[\omega_p^2(x)+\omega_b^2(x)-\omega^2\right]}{\omega_b^2(x)-\omega^2}} \tag{2.15}$$

in the $x_0 \leq x \leq \infty$ half space, i.e. in the medium "2". Applying the W.K.B. condition $S\cong$constant, i.e.

$$E_{z0}^{+}(x,\omega)\sim\frac{A_0^{+}(\omega)}{\sqrt{k_1(x,\omega)}}\sqrt{k_0(\omega)}. \tag{2.16}$$

Therefore the solution is:

$$E_{2zw}(x,t)=$$
$$=\frac{1}{2\pi}\int_{-\infty}^{+\infty}A_0^{+}(\omega)\sqrt[4]{\frac{\omega^2\left[\omega_b^2(x)-\omega^2\right]}{\omega_b(x)\omega_p^2(x)\omega+\omega^2\left[\omega_p^2(x)+\omega_b^2(x)-\omega^2\right]}}\cdot \mathrm{e}^{\mathrm{j}\left[\omega t-\int_{x0}^{x}k_1(\xi,\omega)\mathrm{d}\xi\right]}\mathrm{d}\omega,$$
$$H_{2yw}(x,t)=$$
$$=-\frac{1}{2\pi Z_0}\int_{-\infty}^{+\infty}A_0^{+}(\omega)\sqrt[4]{\frac{\omega_b(x)\omega_p^2(x)\omega+\omega^2\left[\omega_p^2(x)+\omega_b^2(x)-\omega^2\right]}{\omega^2\left[\omega_b^2(x)-\omega^2\right]}}\cdot \mathrm{e}^{\mathrm{j}\left[\omega t-\int_{x0}^{x}k_1(\xi,\omega)\mathrm{d}\xi\right]}\mathrm{d}\omega,$$
$$\tag{2.17}$$

and the signal rotates, i.e.

$$E_{2yw}(x,t)=+\mathrm{j}E_{2zw}(x,t)\ \text{ and }\ H_{2zw}(x,t)=-\mathrm{j}H_{2yw}(x,t). \tag{2.18}$$

We can obtain the other existing modes in half-space"2" in the same manner.

2.1.3 The quasi-homogeneous solution after mode-coupling at the boundary surface A_1

Let this (2.16) signal be generated by an $J_0(x,t)$ current density within $0 \le x \le x_0$ and $0 \le t \le t_0$ as in the homogeneous case (see equation (1.45) in point 1.3), the excitation enters from half-space "1" into the half-space "2" , and the medium in half-space "1" is homogeneous and isotropic with a permittivity ε_0 and permeability μ_0. The arising signal is known, and is described by equations (1.51) in point 1.3.3.

The coupling of signals generated in half spaces "1" and "2" is to be determined in the same way as was seen in homogeneous case (point 1.3.5). The result is a signal, propagating longitudinally in quasi-homogeneous, weakly inhomogeneous plasma generated by an arbitrary $J_0(x,t)$ current impulse – see equation (1.67).

Further, it is known that $k_0(\omega) = \omega/c$ and $k_1(x,\omega)$ is defined by equation (2.15). Now we can get the final form of the signal propagating in half-space "2" in the same way as was seen in points 1.2.5, and the results are

$$E_{2zw}(x,t) = -\frac{Z_0}{4\pi}\int_{-\infty}^{+\infty} I_{x0}(\omega)\sqrt{\frac{k_1(x_0,\omega)}{k_1(x,\omega)}}\frac{k_0(\omega)}{k_0(\omega)+k_1(x_0,\omega)}\cdot e^{j\left[\omega t-\int_{x_0}^{x}k_1(\xi,\omega)d\xi\right]}d\omega,$$

$$H_{2yw}(x,t) = \frac{1}{4\pi}\int_{-\infty}^{+\infty} I_{x0}(\omega)\sqrt{\frac{k_1(x,\omega)}{k_1(x_0,\omega)}}\frac{k_1(x_0,\omega)}{k_0(\omega)+k_1(x_0,\omega)}\cdot e^{j\left[\omega t-\int_{x_0}^{x}k_1(\xi,\omega)d\xi\right]}d\omega,$$

(2.19)

which is the quasi-homogeneous (whistler) solution in frequency band $0 \le \omega \le \omega_b$. In practical computation we shall use the real space-time function forms of solutions (2.19), which are analogous to equation (1.70).

2.1.4 Coupled W.K.B. solution in half-space "2"

In point 2.1.2 we got the quasi-homogeneous solution in half-space "2" applying the W.K.B. approximation in a generalised form for signals with arbitrary shape. If the inhomogeneity is stronger than in a quasi-homogeneous case, we can use coupled W.K.B. approximation (*Budden,* 1966; *Ferencz,* 1996). It can be seen from the precedents (point

2.1.2) that to find the coupled W.K.B. solution for the inhomogeneous case one may start from equation (2.14) and (2.17), i.e.

$$E^+_{2zw}(x,\omega)=A^+_0(\omega)\sqrt{\frac{k_0(\omega)}{k_1(x,\omega)}}\cdot \mathrm{e}^{-\mathrm{j}\int_{x0}^{x}k_1(\xi,\omega)d\xi},$$
$$H^+_{2yw}(x,\omega)=-A^+_0(\omega)\frac{1}{Z_0}\sqrt{\frac{k_1(x,\omega)}{k_0(\omega)}}\cdot \mathrm{e}^{-\mathrm{j}\int_{x0}^{x}k_1(\xi,\omega)d\xi}, \tag{2.20}$$

and the backward propagating (reflected or backscattered) modes

$$E^-_{2zw}(x,\omega)=A^-_0(\omega)\sqrt{\frac{k_0(\omega)}{k_1(x,\omega)}}\cdot \mathrm{e}^{+\mathrm{j}\int_{x0}^{x}k_1(\xi,\omega)d\xi},$$
$$H^-_{2yw}(x,\omega)=A^-_0(\omega)\frac{1}{Z_0}\sqrt{\frac{k_1(x,\omega)}{k_0(\omega)}}\cdot \mathrm{e}^{+\mathrm{j}\int_{x0}^{x}k_1(\xi,\omega)d\xi}. \tag{2.21}$$

The complete solution in any space-point is given by

$$E_{2zw}=E^+_{2zw}+E^-_{2zw} \quad \text{and} \quad H_{2yw}=H^+_{2yw}+H^-_{2yw}. \tag{2.22}$$

Substituting these expressions into Maxwell's equations

$$\frac{\partial H_y}{\partial x}=-\frac{k_1^2(x,\omega)}{\mathrm{j}\mu_0\omega}\cdot E_z \quad \text{and} \quad \frac{\partial E_z}{\partial x}=\mathrm{j}\mu_0\omega\cdot H_y,$$

and executing the normal calculations the coupled W.K.B. equations will be obtained, which modify the (2.20)~(2.22) relations by replacing

$A_0^{\pm}(\omega)$ by $A_0^{\pm}(x,\omega)$.

The coupling equations are now

$$\frac{\partial A^+_0(x,\omega)}{\partial x}=\frac{1}{2}\frac{\partial \ln[k_1(x,\omega)]}{\partial x}\cdot A^-_0(x,\omega)\cdot \mathrm{e}^{\mathrm{j}2\int_{x0}^{x}k_1(\xi,\omega)d\xi},$$
$$\frac{\partial A^-_0(x,\omega)}{\partial x}=\frac{1}{2}\frac{\partial \ln[k_1(x,\omega)]}{\partial x}\cdot A^+_0(x,\omega)\cdot \mathrm{e}^{-\mathrm{j}2\int_{x0}^{x}k_1(\xi,\omega)d\xi}. \tag{2.23}$$

In numerical computations the solutions of equation-system (2.23) may be obtained using successive approximation. In this approximation the starting values of $A_0^{\pm}$ are

$$A_0^+(x,\omega) = \text{constant}(\omega) \quad \text{and} \quad A_0^-(x,\omega) = 0.$$

In the next step the second equation of (2.23) gives the first approximation of $A_0^-(x,\omega)$ and applying this function in the first equation of (2.23) we can get the first correction of $A_0^+(x,\omega)$. The procedure is convergent in most cases of inhomogeneities. However, one may note that simple approximate solutions cannot easily be obtained by equation-system (2.23).

2.1.5 Solution for strong inhomogeneities (in half-space "2") using only one space-variable

In the present model (see Fig. 1.6 and the earlier parts of point 2.1) the medium in half-space "2" is inhomogeneous in the x-direction but remains independent of space-variable y and z. It is known (*Osterberg*, 1958; *Heald and Wahrton*, 1965) that in inhomogeneous media varying only in a single direction the solution may be sought in general form by applying the generalised propagation vector (see more generally in details e.g. *Ferencz* 1977a,b, 1978a, 1987).

Let us assume – based upon our earlier results – that the sought, longitudinally propagating, plane wave (whistler-mode) solution is a rotating signal, see e.g. equation (2.17). Using these assumptions the equation-system (1.29) can be rewritten, than applying Laplace-transform, finally an $s \rightarrow j\omega$ transformation and applying (2.15), the following equations will be obtained

$$\begin{aligned} \frac{\partial H_y(x,\omega)}{\partial x} &= j\frac{k_1^2(x,\omega)}{\mu_0\omega}\cdot E_z(x,\omega), \\ \frac{\partial E_z(x,\omega)}{\partial x} &= j\mu_0\omega\cdot H_y(x,\omega). \end{aligned} \tag{2.24}$$

It can be derived from (2.24) that

$$\frac{\partial^2 E_z(x,\omega)}{\partial x^2} = -k_1^2(x,\omega)\cdot E_z(x,\omega). \tag{2.25}$$

Let the solution be searched in the form of

$$E_z(x,\omega) = E_{z0}(\omega) \cdot e^{-j\int_{x_0}^{x} K(\xi,\omega)d\xi}, \tag{2.26}$$

where $K(x, \omega)$ is the generalised, complex propagation-factor. Applying (2.26) in equation (2.25) we get a new form of equation in which the unknown is the $K(x, \omega)$ function

$$K^2(x,\omega) + j\frac{\partial K(x,\omega)}{\partial x} = k_1^2(x,\omega). \tag{2.27}$$

Equation (2.27) is a known class of differential equations and can be solved by using known methods (*Kamke,* 1956). The solution may be derived from equation (2.25) or from equation (2.27), respectively. For example the general mathematical form of the equation (2.25) is

$$y'' = f(x) \cdot y, \tag{2.28}$$

and the solution of this equation is

$$y(x) = C \cdot e^{\int u(x)dx}, \tag{2.29}$$

where $u(x) = y'/y$ and $u(x)$ is the solution of the equation

$$u' + u^2 = f(x), \tag{2.30}$$

where this equation is identical to equation (2.27) if

$$f(x) = -k_1^2(x,\omega). \tag{2.31}$$

Using successive approximations it is possible to find the solution with prescribed accuracy. Let the first approximation be

$$u_1^2(x) \cong -k_1^2(x,\omega), \tag{2.32}$$

which is the known solution of the quasi-homogeneous case. Therefore the next step is

$$\frac{\partial u_1(x)}{\partial x} = \pm j\frac{\partial k_1(x,\omega)}{\partial x} \quad \text{etc.} \tag{2.33}$$

or in a general form

$$u_{i+1}^2(x,\omega) \cong -k_1^2(x,\omega) - \frac{\partial u_i(x,\omega)}{\partial x}. \tag{2.34}$$

Using these approximations we get more and more accurate formulas. For example, the second approximation is

$$u_2(x,\omega) = \pm \mathrm{j}k_1(x,\omega) - \frac{1}{2}\frac{1}{k_1(x,\omega)}\frac{\partial k_1(x,\omega)}{\partial x}, \tag{2.35}$$

and this gives exactly the solution (2.19). After two new approximate steps one arrives at the next approximate solution, already with the first approximation of attenuation caused by the backscattering generated by the inhomogeneity of medium

$$E_{2zw}(x,\omega) \cong \frac{-Z_0}{4\pi}\int_{-\infty}^{+\infty} I_{x0}(\omega)\sqrt{\frac{k_1(x_0,\omega)}{k_1(x,\omega)}}\frac{k_0(\omega)}{k_0(\omega)+k_1(x_0,\omega)}\mathrm{e}^{-\int_{x0}^{x}\alpha_4(\xi,\omega)\mathrm{d}\xi}\,\mathrm{e}^{\mathrm{j}\left[\omega t - \int_{x0}^{x} K_3(\xi,\omega)\mathrm{d}\xi\right]}\mathrm{d}\omega \tag{2.36}$$

where

$$K_3(x,\omega) \cong k_1(x,\omega)\left\{1 - \frac{1}{4k_1^4(x,\omega)}\left[k_1(x,\omega)\frac{\partial^2 k_1(x,\omega)}{\partial x^2} - \left(\frac{\partial k_1(x,\omega)}{\partial x}\right)^2\right]\right\} \tag{2.36a}$$

and

$$\alpha_4(x,\omega) \cong \frac{1}{2}\frac{\partial[\ln k_1(x,\omega)]}{\partial x}\left\{\frac{1}{k_1^3(x,\omega)}\frac{\partial^2 k_1(x,\omega)}{\partial x^2} - \right.$$

$$\left. - \frac{3}{4k_1^4(x,\omega)}\left[\frac{\partial k_1(x,\omega)}{\partial x}\right]^2 - \frac{1}{4k_1^2(x,\omega)}\left[\frac{\partial k_1(x,\omega)}{\partial x}\right]^{-1}\frac{\partial^3 k_1(x,\omega)}{\partial x^3}\right\}. \tag{2.36b}$$

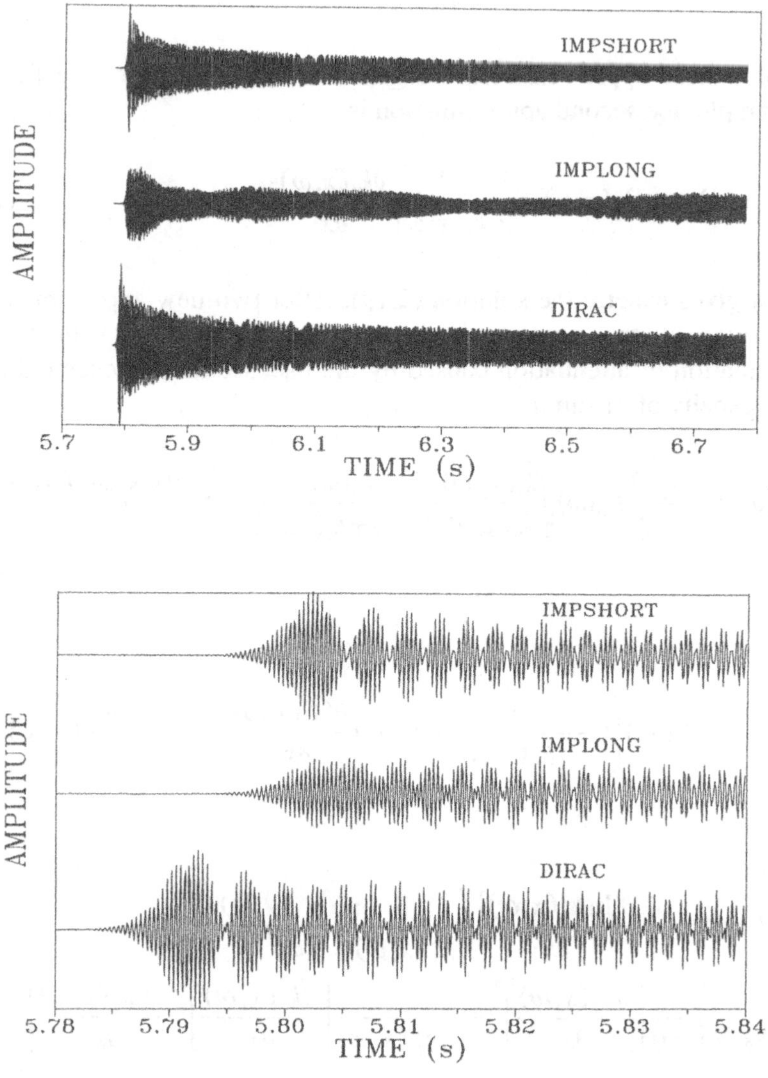

Figure 2.1 Calculated whistlers of Dirac-delta (DIRAC), short (IMPSHORT) and long (IMPLONG) impulse excitations if $L=5$ and $N_1=10000$ cm^{-3}, where N_1 is electron density at 1000 km height above the Earth-surface.

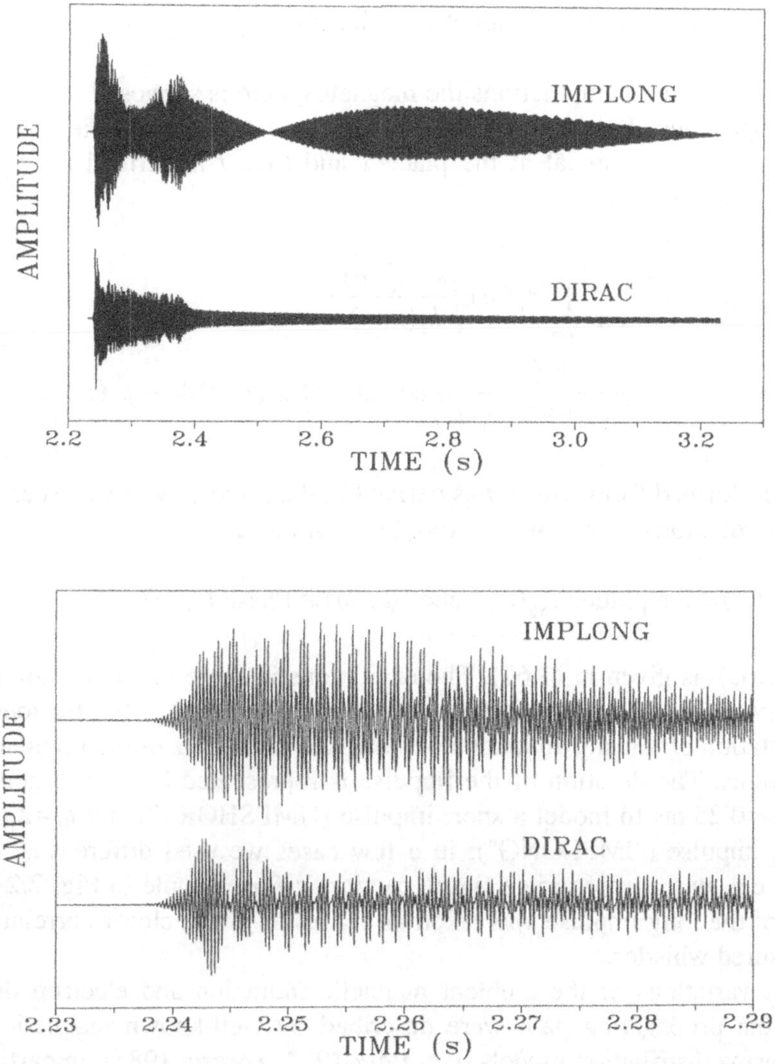

Figure 2.2 Calculated whistlers of Dirac-delta (DIRAC) and long impulse (IMPLONG) excitations if L=4.654 and N_1=2550 cm^{-3}. These whistlers are comparable directly to measured (at Halley) and analysed whistlers (see Fig. 4.18 or *Hamar et al.,* 1992).

The correctional terms are small in the case of the Earth's plasmasphere, as is clear from equations (2.36). Therefore, our expectation is that the approximation (2.19) will be usable in most cases of investigation of the near-Earth space, but we must use the *more precise solutions in some regions of the Solar System,* or in other wave propagational problems.

2.1.6 Some results of model-calculations

In the present computations the magnetosphere is supposed to be weakly inhomogeneous. It is known that in this case the electric field intensity component of the signal at the place x and time t is defined by equation (2.19) and therefore

$$E_{2zw}(x,t) \cong -\frac{Z_0}{2\pi}\int_{\omega_{\min}}^{\omega_{\max}} A_I(\omega)\sqrt{\frac{k_1(x_0,\omega)}{k_1(x,\omega)}}\cdot \frac{k_0(\omega)}{k_0(\omega)+k_1(x_0,\omega)}\cos\left[\omega t-\int_{x_0}^{x} k_1(\xi,\omega)\mathrm{d}\xi+\varphi_I(\omega)\right]\mathrm{d}\omega \tag{2.37}$$

in the re-formed form which was derived in the same way as was seen in the case of equations (1.69)→(1.70) too. In equation (2.37)

$$A_I(\omega) \equiv \text{Amplitude } I_{x0}(\omega) \quad \text{and} \quad \varphi_I(\omega) \equiv \text{Phase } I_{x0}(\omega),$$

and $I_{x0}(\omega)$ is given in (1.67). The excitations in these model computations are given in point 1.3.6 (see equations (1.71)~(1.74)). Besides the computation of transfer functions we shall see the effects of the impulse-type excitations. The duration of the impulse is represented by t_0. We generally took t_0=0.25 ms to model a short impulse ("IMPSHORT") and t_0=2 ms for a long impulse ("IMPLONG"); in a few cases we used different t_0 values (noted e.g. in the appropriate figure captions); for example in Fig. 2.2 the t_0 value of the long impulse was taken as 1.2 ms to give a closer correlation to a measured whistler.

The variations of the ambient magnetic induction and electron density along the propagating path were described by well-known magnetic field and plasma distribution models (e.g. *Park* 1972; *Tarcsai* 1981). In particular, we used the usual dipole magnetic field model $B_{F0}(x)$ and *Park's* (1972) diffusive equilibrium model DE-1 for $N(x)$, where x is the arc-length along the magnetic line of force measured from a height of 1000 km above the Earth's surface on the source hemisphere, i.e. on the hemisphere of the source signal $\overline{J}_1$. This allowed valid comparisons to be made between calculated and observed signals (whistlers, i.e. electron whistlers).

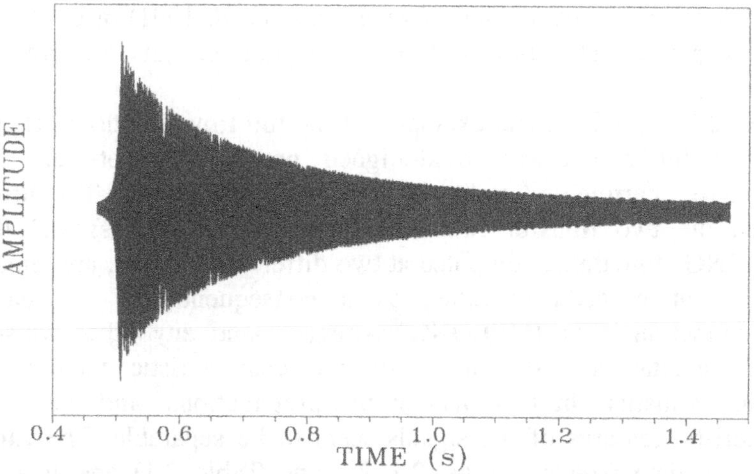

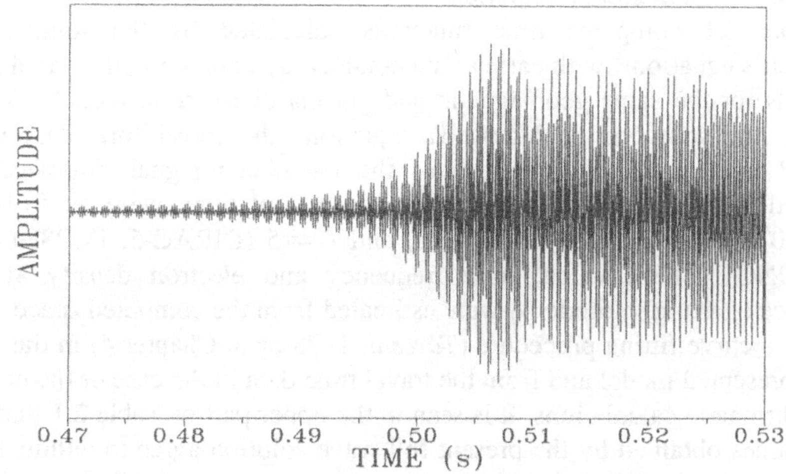

Figure 2.3 Calculated whistler of Dirac-delta excitation (i.e. transfer function) if L=2.2 and N_1=10000 cm^{-3}.

The time function of the whistler waveform was calculated at a given point, (corresponding to the experimental data); then this calculated signal was analysed by Fast Fourier Transformation (FFT) and by matched filtering

(*Hamar and Tarcsai*, 1982, 1991 in the same way, as in the case of measured signals); finally the results of the analysis of calculated and observed whistlers, the latter recorded at Halley Station of British Antarctic Survey, Antarctica, or on board of the ACTIVE (Intercosmos-24) satellite (by the SAS experiment – part 3.2 and *Lichtenberger et al.* 1991) or other satellites were compared. (The used methods of signal analysis are presented in Chapter 4.)

Figs. 2.1 and 2.2 present examples of the full time functions. These were computed for a one-way field-aligned propagation between the two hemispheres, starting and ending at an altitude of 1000 km. The differences between the two transfer functions ("DIRAC" responses) and the two "IMPLONG" functions, computed at two different L-values, are remarkable. However, at a given L-value, as a consequence of the nature of transfer-function, the "IMPLONG" whistler –and any other whistlers of arbitrary excitation – also manifests the characteristic features of the "DIRAC" whistler. In this manner the propagational and the excitation characteristic features of the signals seem to be separable. The calculated travel time data (see e.g. Figs. 2.1~2.3 and Table 2.1) are in a general agreement with those computed earlier using monochromatic solutions for the propagation in this simple case and computing the travel time on a given frequency by numerical integration.

Table 2.1 compares time functions calculated by the solution of Maxwell's equations, with earlier data obtained by computing the travel time integrals for the same geomagnetic and plasma distribution models. In the case of the solutions of Maxwell's equations the travel time data were derived by using FFT spectrum of the calculated signal, followed by matched filtering. The input data for the calculations were as follows: $N_1 = 10^4\,\text{cm}^{-3}$, $L = 2.2$ (DIRAC-2.2) and $L = 5$ (DIRAC-5, IMPSHORT, IMPLONG). The L-value, nose frequency and electron density at the reference (starting) altitude N_1 were estimated from the computed space-time data by a curve fitting procedure (*Tarcsai,* 1975 or in Chapter 4) in the case of the presented model and from the travel time data in the case of the earlier monochromatic calculations. It is seen in the upper part of Table 2.1 that the travel times obtained by the present full wave solution agree to within 1 ms with those obtained by numerical integration presenting smaller systematic differences too. From the excellent agreement of travel times it follows that the curve fitting procedure applied for these simulated whistlers yields very similar results for both types of data (lower part of Table 2.1). It is also seen that the scaled travel times of the whistler excited by a longer impulse tend to be (slightly) larger than those generated by a DIRAC-impulse and this

Table 2.1 Comparison of travel times at various frequencies, and other whistler parameters, at $L = 2.2$ and $L = 5$, calculated from the solution of Maxwell's equations and by the conventional, strictly monochromatic integration along a field line.

	Solution of Maxwell's equations				**Numerical integration**	
	DIRAC-2.2	DIRAC-5	IMPSHORT	IMPLONG	MODELL-2.2	MODELL-5
Frequency (Hz)	Travel-time (s)					
1500	0.833	6.317	6.318	6.319	0.833	6.318
2000	0.726	5.931	5.931	5.932	0.727	5.931
2500	0.653	5.802	5.802	5.802	0.654	5.801
3000	0.600	5.857	5.858	5.859	0.601	5.857
3500	0.559	6.083	6.084	6.085	0.560	6.083
4000	0.527	6.505	6.506	6.507	0.528	6.505
Results of curve fitting (see the method in Chapter 4)						
L-value	2.205	4.999	4.999	4.999	2.211	4.999
Nose frequency (Hz)	29350	2590	2590	2590	29129	2590
N_1 (cm^{-3})	8780	9870	9950	9920	9400	9910

phenomenon can not be simulated by any monochromatic or quasi-monochromatic descriptions. That is, the scaled travel times depend definitely on the form of the excitation spectrum. This comparison verifies the accuracy and usefulness of the presented new solution of Maxwell's equations.

Fig. 2.4 shows a real whistler recorded on board of the Intercosmos-5 satellite (*Mihajlova,* 1977; *Hamar and Tarcsai,* 1991). Figs. 2.5 and 2.6 show enlarged portions of calculated (DIRAC) whistlers – transfer functions – presented in Figs. 2.2 and 2.3, respectively. It is clear that the characteristic signal forms correspond to each other; the whistler has a complex waveform and it is not a simple result of interference of the two "sinusoidal" waves of varying frequency.

Furthermore matched filtering (see in point 4.1 and in *Hamar and Tarcsai,* 1982, 1991; *Hamar et al.* 1992) was applied by us to all the calculated whistlers in the same manner as in the case of measured signals. The amplitude patterns (*Tarcsai et al.* 1989; *Hamar et al.* 1990) of the time functions seen in Fig. 2.1 (L = 5, N_l = 10000 cm^{-3}) are presented in Fig. 2.7. This – besides the known accuracy problem at the nose frequency of the matched filtering technique as presently implemented (parts 4.1 and 4.2) – shows examples of the effect of excitation spectrum $I_{x0}(\omega)$ entering the magnetosphere. Note that the whistler spectrum decreases periodically in the case of a long impulse type excitation ("IMPLONG") with a periodicity $1/t_0$; in this example $1/t_0 = 500\,\text{Hz}$. This periodicity is 4 kHz in the case of the computed "IMPSHORT" signal.

The general character of the amplitude distribution of a whistler measured at Halley Station (British Antarctic Survey) (Fig. 2.8) correlates with the pattern of the DIRAC whistler calculated with the same L and N_l. However, the spacing of the amplitude minima and maxima (cf. Fig. 2.7) suggest more complex excitation, though further investigation is needed to assess whether this amplitude behaviour can arise for other reasons (see points 2.2 and 4.4). These first results suggest the possibility of inferring the form of the exciting signal by applying e.g. the matched filtering method in conjunction with the whistler mode solution of Maxwell's equations.

A very detailed analysis has already been made of the whistlers recorded at Halley and marked "A" in Fig. 2.9 (*Hamar et al.,* 1990); the parameters of the propagation path are L = 4.654 and N_l = 2550 cm^{-3}. Applying these parameters we calculated two whistlers: one excited by a DIRAC-delta and the other by a long impulse (t_0 = 1.2 ms). The FFT spectra of these modelled whistlers are shown in Figs. 2.10 and 2.11, respectively.

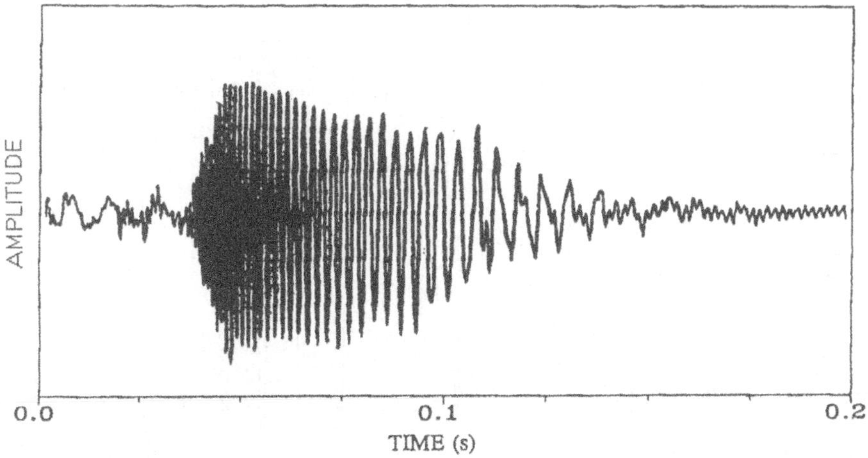

Figure 2.4 A whistler measured onboard of the Interkosmos-5 satellite (after G.A. Mihajlova, - *Mihajlova*, 1977; *Hamar and Tarcsai*, 1991; *Ferencz et al.*, 1996)

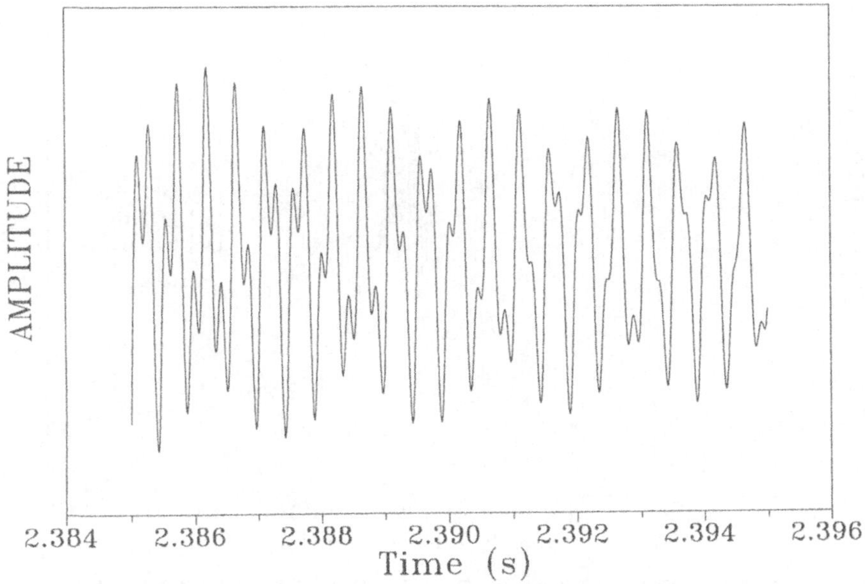

Figure 2.5 An enlarged section of Fig. 2.2. DIRAC-whistler (transfer function).

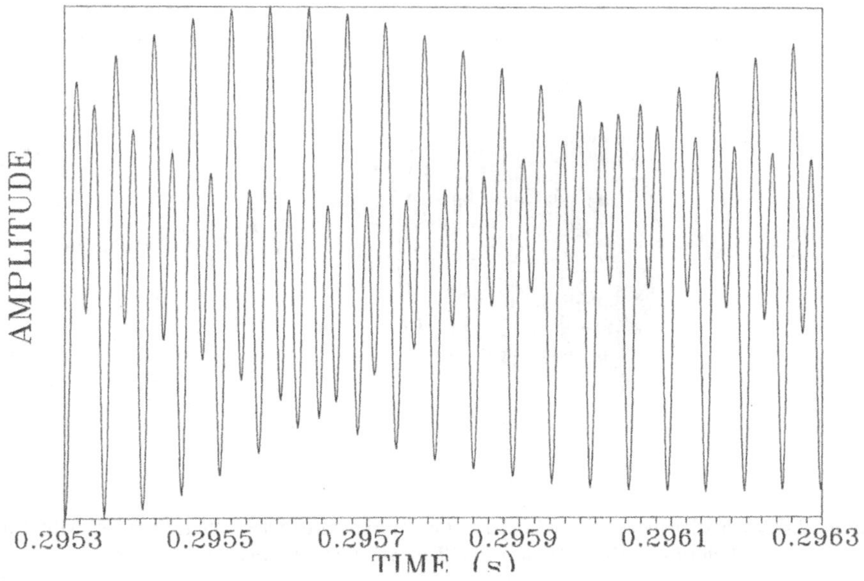

Figure 2.6 An enlarged section of Fig. 2.3. whistler (transfer function).

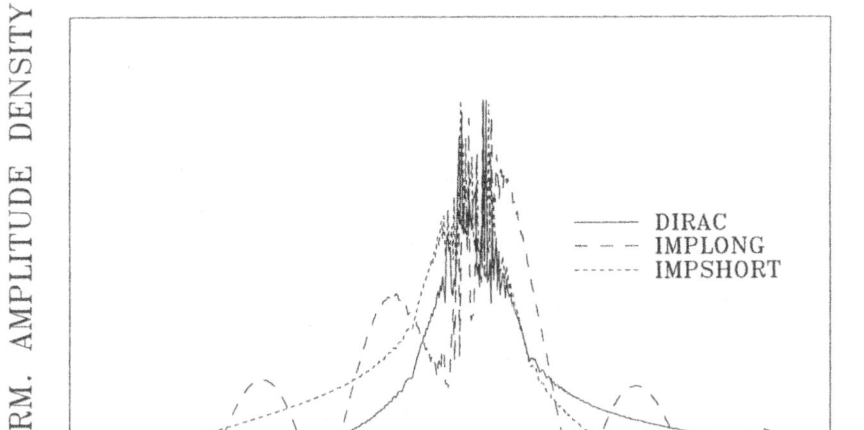

Figure 2.7 The amplitude distributions of calculated whistler presented in Fig. 2.1 after matched filtering.

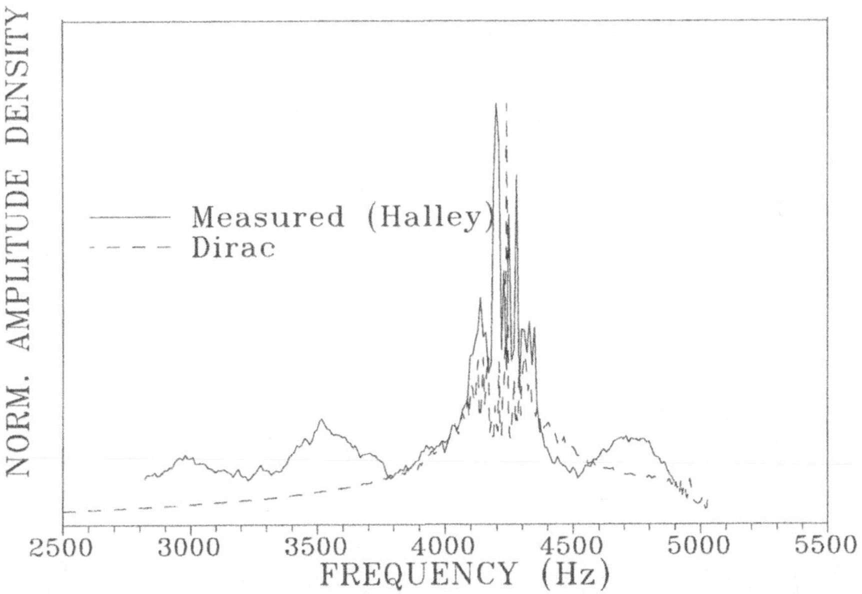

Figure 2.8 The amplitude distributions of a measured whistler (at Halley, 9 August 1989 16:16:56 UT; courtesy of dr. A. J. Smith, British Antarctic Survey, Cambridge, U.K.) and the transfer function, where L=4.276 and N_1=5780 cm^{-3}.

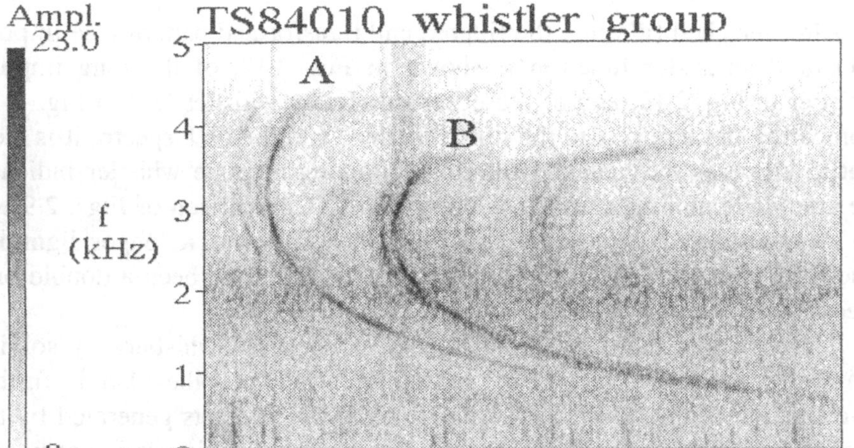

Figure 2.9 The FFT spectrum of a whistler measured at Halley on 21 July 1984, 17:14:14 UT – courtesy of dr. A. J. Smith, British Antarctic Survey, Cambridge, U.K.–, where L=4.654 and N_1=2550 cm^{-3}.

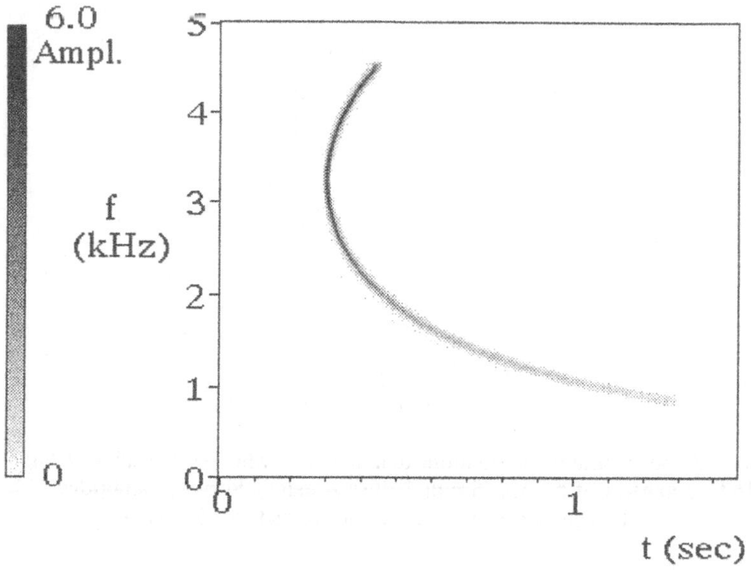

Figure 2.10 The FFT spectrum of computed transfer function if L=4.654, N_1=2550 cm^{-3} and ω_{max}=4.5 kHz.

After matched filtering, the filtered and transformed pattern (see in point 4.1) of the transfer function is shown in Fig. 2.12, of the long impulse whistler in Fig. 2.13, and of the observed Halley whistler "A" in Fig. 2.14. Comparing these results to the time functions and the FFT spectra it is clear that in all cases the matched filtering following a single whistler indicates the amplitude minima as trace splitting events. Comparison of Figs. 2.9 and 2.11 suggests that the excitation of whistler "A" was a longer lightning discharge (spheric); alternatively, the whistler may have been a double one. (See more about this item later in point 4.4.)

It can be said that the modelling of the observed whistlers by solving Maxwell's equations has been successful. On the other hand, further investigation of the effects of excitations and the artefacts generated by the applied analytical methods of measured and computed signals is necessary; and this new full-wave solution is a good tool in solving this problem. (See as a demonstration of this new way Fig. 2.15).

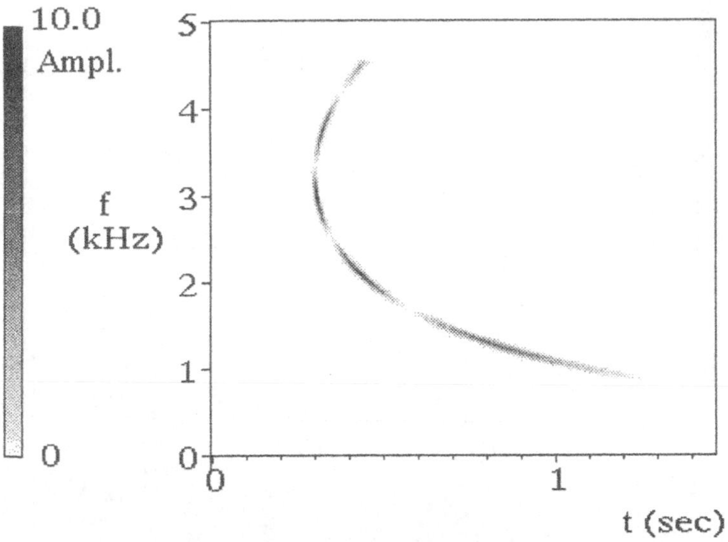

Figure 2.11 The FFT spectrum of computed IMPLONG-whistler if L=4.654, N_1=2550 cm^{-3}, t_0=1.2 ms and ω_{max}=4.5 kHz.

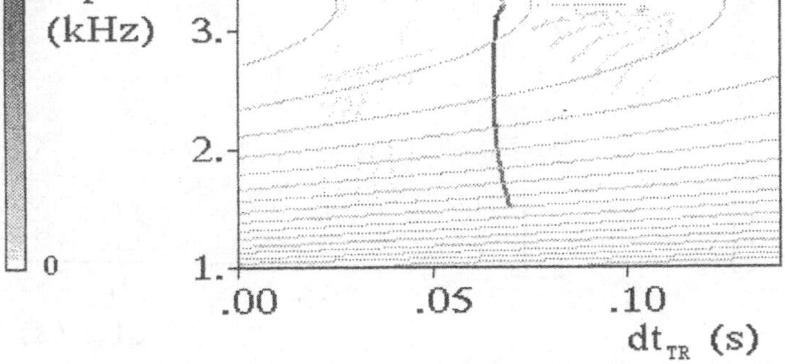

Figure 2.12 The pattern of transfer function presented in Fig. 2.10 after matched filtering.

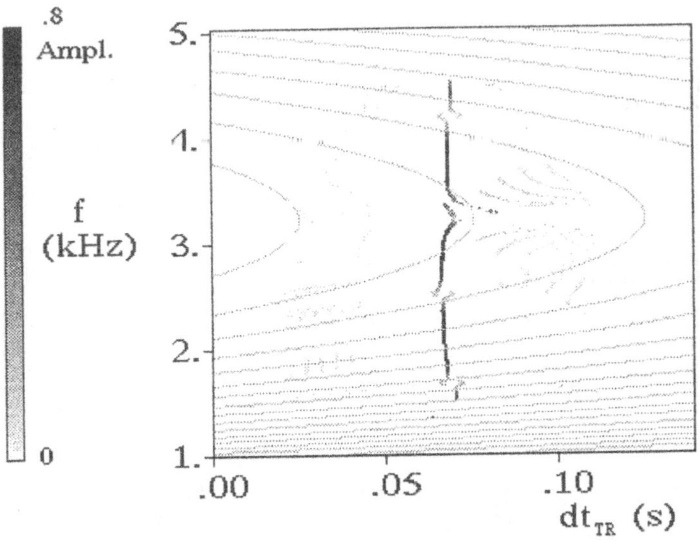

Figure 2.13 The pattern of the IMPLONG-whistler presented in Fig. 2.11 after matched filtering.

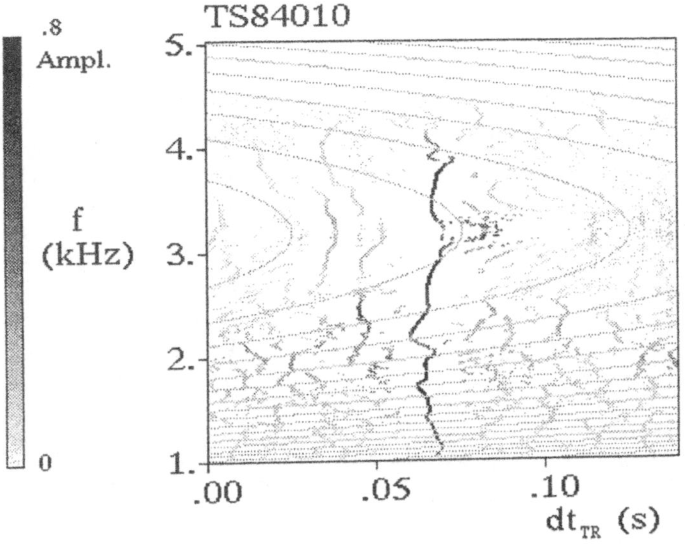

Figure 2.14 The pattern of the measured whistler presented in Fig. 2.9 after matched filtering.

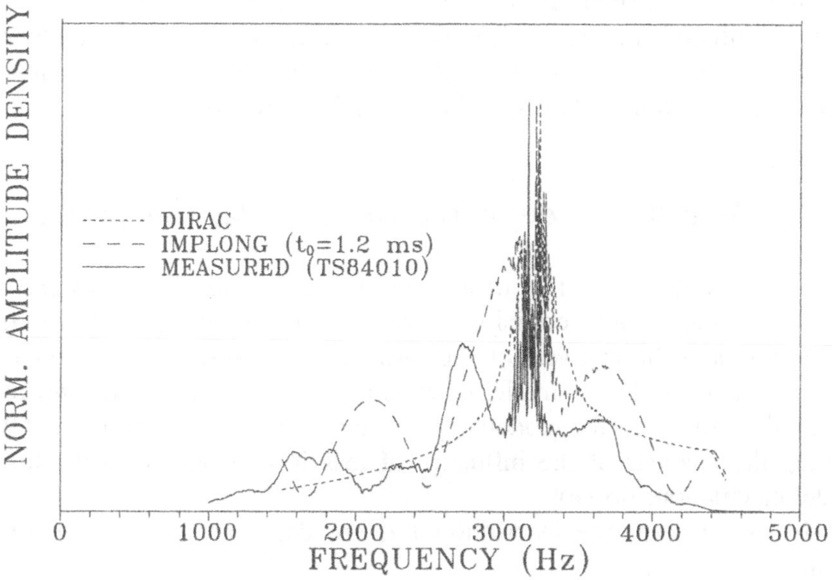

Figure 2.15 The amplitude distributions of signals presented in Figs. 2.12~2.14.

2.1.7 Conclusions

We succeeded in finding the general whistler-mode solution even for an *inhomogeneous*, cold, anisotropic magnetoionic medium in the case of strictly longitudinal propagation. Using this solution it is possible to find more accurate VLF propagational models to investigate the fine-structure of signals, to find the causes of this structures, to find the character of source signals of specific formed propagating signals and to develop advanced signal analysing (e.g. advanced matched filtering) techniques.

It seems that the application of this real full-wave solution of Maxwell's equations means a real development in the interpretation of the characteristic features of observed signals. Besides this, the calculated signals give an opportunity to test the detailed nature and accuracy of presently-used advanced whistler analysis methods (i.e. matched filtering, parameter estimation), as regards, for example, filtering at the vicinity of the nose, analysing double and multiple whistlers, filtering signals of a wide range of excitation spectra.

As could have been expected, the transfer function is especially useful in discriminating between the propagational and excitation effects. This gives the possibility of interpreting features that are at present not well understood. Finally, one may suggest replacing the old approximate whistler models by a new one in the commonly used whistler analysis methods.

2.2 Whistler-mode solution in lossy electronplasma

In this chapter, the effect of loss will be taken into account by the well-known particle-particle collision frequency (ν_c) in the plasma model. It makes possible the analysis of the influence of loss in comparison to the lossless – collisionless – model-calculations. These calculations and results extend the validity of the models – presented in details in parts 1.3 and 2.1 –, making the analysis of the influence of loss in comparison to the lossless model-calculations possible.

The structure and the description of the model can be seen in Fig. 1.6 and in part 1.3.

2.2.1 Solution if the medium "2" is a homogeneous plasma

As a starting point, the magnetised plasma is supposed to be homogeneous, the form of Maxwellian equation-system is concordant with the one for vacuum and the plasma parameters are described by current density vector, see equation (1.24). (The meaning of symbols can also be seen in Chapter 1.) However, this equation-system has some changing. The interaction between the signal and the medium is specified by equations (1.25) containing the presence of loss in the plasma

$$
\begin{aligned}
& m\frac{\partial \overline{v}}{\partial t} + \overline{F}_{\text{loss}} = q\left(\overline{E} + \overline{v} \times \overline{B}_{F0}\right), \\
& \overline{F}_{\text{loss}} = m\nu_c \overline{v}, \\
& \overline{J} = qN\overline{v}, \\
& \overline{\nabla}\overline{J} + \frac{\partial \rho}{\partial t} = 0.
\end{aligned}
\qquad \textbf{(2.38)}
$$

We use the common notations of plasma- and gyro-frequency as these were defined earlier. The frequency band of the signal searched for is $0 < \omega < \omega_b$

(whistler mode) during the interpretation of results. The differential equation describing the relation between the electromagnetic signal to be determined and the parameters of the plasma, is as follows

$$m\frac{\partial \overline{v}}{\partial t}+m\nu_c\overline{v}=q\left(\overline{E}+\overline{v}\times\overline{B}_{F0}\right). \tag{2.39}$$

Rearranging (2.39), the following category of differential equations will be recognisable:

$$\frac{\partial \overline{v}}{\partial t}+\nu_c\overline{v}-\left(-\frac{q}{m}B_{F0}\right)\left(\overline{e}_x\times\overline{v}\right)=\frac{q}{m}\overline{E}. \tag{2.40}$$

In general form (2.40) is identical to (1.26a), however, the component equations of (2.40) are different from (1.26b). They are

$$\begin{aligned}&\frac{\partial v_x}{\partial t}+\nu_c v_x=f_x(x),\\&\frac{\partial v_y}{\partial t}+\omega_b v_z+\nu_c v_y=f_y(x),\\&\frac{\partial v_z}{\partial t}-\omega_b v_y+\nu_c v_z=f_z(x).\end{aligned} \tag{2.41}$$

The complete solutions of (2.41) – in accordance with (*Kamke,* 1956, 1965) – are as follows

$$\begin{aligned}&v_x=\mathrm{e}^{-\nu_c t}\left[c+\int^{t}\frac{q}{m}E_x(\tau)\mathrm{e}^{\nu_c\tau}\,\mathrm{d}\tau\right],\\&v_y=\mathrm{e}^{-\nu_c t}\left(c_1\cos\omega_b t+c_2\sin\omega_b t\right)+\frac{1}{\omega_b}\int^{t}\varphi_y(\tau)\mathrm{e}^{-\nu_c(t-\tau)}\sin\omega_b(t-\tau)\mathrm{d}\tau,\\&v_z=\mathrm{e}^{-\nu_c t}\left(c_1\cos\omega_b t+c_2\sin\omega_b t\right)+\frac{1}{\omega_b}\int^{t}\varphi_z(\tau)\mathrm{e}^{-\nu_c(t-\tau)}\sin\omega_b(t-\tau)\mathrm{d}\tau,\end{aligned} \tag{2.42}$$

where

$$\varphi_y(\tau)=\frac{q}{m}\left[E_y'(\tau)+\nu_c\, E_y(\tau)-\omega_b\, E_z(\tau)\right],$$

$$\varphi_z(\tau)=\frac{q}{m}\left[E_z'(\tau)+\nu_c\, E_z(\tau)+\omega_b\, E_y(\tau)\right].$$

The homogeneous terms do not propagate, as they have no connection to the excitation and they are not real signals, therefore these parts will be eliminated.

Writing the velocity components back into (2.38), the components of the current density vector can be obtained as follows

$$J_x = qNv_x = \varepsilon_0\omega_p^2 \mathrm{e}^{-\nu_c t}\cdot\left[\int^t E_x(\bar{r},\tau)\cdot \mathrm{e}^{\nu_c\tau}\mathrm{d}\tau\right],$$

$$J_y = \varepsilon_0\omega_p^2\left\{\frac{1}{\omega_b}\int^t \frac{\partial E_y}{\partial\tau}\mathrm{e}^{-\nu_c(t-\tau)}\sin\omega_b(t-\tau)\mathrm{d}\tau + \right.$$

$$+\frac{1}{\omega_b}\nu_c\int^t E_y \mathrm{e}^{-\nu_c(t-\tau)}\sin\omega_b(t-\tau)\mathrm{d}\tau -$$

$$\left. -\int^t E_z \mathrm{e}^{-\nu_c(t-\tau)}\sin\omega_b(t-\tau)\mathrm{d}\tau\right\}, \qquad \mathbf{(2.43)}$$

$$J_z = \varepsilon_0\omega_p^2\left\{\frac{1}{\omega_b}\int^t \frac{\partial E_z}{\partial\tau}\mathrm{e}^{-\nu_c(t-\tau)}\sin\omega_b(t-\tau)\mathrm{d}\tau + \right.$$

$$+\frac{1}{\omega_b}\nu_c\int^t E_z \mathrm{e}^{-\nu_c(t-\tau)}\sin\omega_b(t-\tau)\mathrm{d}\tau +$$

$$\left. +\int^t E_y \mathrm{e}^{-\nu_c(t-\tau)}\sin\omega_b(t-\tau)\mathrm{d}\tau\right\}.$$

Substituting (2.43) into (1.24) the following differential-equations system will be obtained

$$
I.\begin{cases}
\dfrac{\partial H_z}{\partial y}-\dfrac{\partial H_y}{\partial z}=\varepsilon_0\left\{\omega_p^2 \mathrm{e}^{-\nu_c t}\int^{t} E_x(\tau)\mathrm{e}^{\nu_c \tau}\mathrm{d}\tau+\dfrac{\partial E_x}{\partial t}\right\}, \\
\dfrac{\partial H_x}{\partial z}-\dfrac{\partial H_z}{\partial x}=\varepsilon_0\left\{\dfrac{\omega_p^2}{\omega_b}\int^{t}\dfrac{\partial E_y}{\partial \tau}\mathrm{e}^{-\nu_c(t-\tau)}\sin\omega_b(t-\tau)\mathrm{d}\tau+\right. \\
\qquad +\dfrac{\omega_p^2}{\omega_b}\nu_c\int^{t} E_y(\tau)\mathrm{e}^{-\nu_c(t-\tau)}\sin\omega_b(t-\tau)\mathrm{d}\tau- \\
\qquad \left.-\omega_p^2\int^{t} E_z(\tau)\mathrm{e}^{-\nu_c(t-\tau)}\sin\omega_b(t-\tau)\mathrm{d}\tau+\dfrac{\partial E_y}{\partial t}\right\}, \\
\dfrac{\partial H_y}{\partial x}-\dfrac{\partial H_x}{\partial y}=\varepsilon_0\left\{\dfrac{\omega_p^2}{\omega_b}\int^{t}\dfrac{\partial E_z}{\partial \tau}\mathrm{e}^{-\nu_c(t-\tau)}\sin\omega_b(t-\tau)\mathrm{d}\tau+\right. \\
\qquad +\dfrac{\omega_p^2}{\omega_b}\nu_c\int^{t} E_z(\tau)\mathrm{e}^{-\nu_c(t-\tau)}\sin\omega_b(t-\tau)\mathrm{d}\tau+ \\
\qquad \left.+\omega_p^2\int^{t} E_y(\tau)\mathrm{e}^{-\nu_c(t-\tau)}\sin\omega_b(t-\tau)\mathrm{d}\tau+\dfrac{\partial E_z}{\partial t}\right\},
\end{cases}
$$

$$
II.\begin{cases}
\dfrac{\partial E_z}{\partial y}-\dfrac{\partial E_y}{\partial z}=-\mu_0\dfrac{\partial H_x}{\partial t}, \\
\dfrac{\partial E_x}{\partial z}-\dfrac{\partial E_z}{\partial x}=-\mu_0\dfrac{\partial H_y}{\partial t}, \\
\dfrac{\partial E_y}{\partial x}-\dfrac{\partial E_x}{\partial y}=-\mu_0\dfrac{\partial H_z}{\partial t}.
\end{cases} \qquad \mathbf{(2.44)}
$$

Substituting group II. into the derived form of group I. according to time it will be clear that

$$\frac{1}{c^2}\left(\omega_p^2 E_x+\frac{\partial^2 E_x}{\partial t^2}\right)=0,$$

and therefore $E_x\equiv 0$, as was presented in the case of equation (1.30a). The supposed waveform in agreement with the excitation is plane-wave propagating to x-direction, so

$$\frac{\partial}{\partial y}=\frac{\partial}{\partial z}\equiv 0.$$

The equation-system to be solved is as follows

$$\frac{\partial^2 E_y}{\partial x^2}=\frac{1}{c^2}\left\{\omega_p^2\int_0^t \frac{\partial E_y}{\partial \tau}e^{-\nu_c(t-\tau)}\cos\omega_b(t-\tau)d\tau+\right.$$
$$+\nu_c\omega_p^2\int_0^t E_y(\tau)e^{-\nu_c(t-\tau)}\cos\omega_b(t-\tau)d\tau-$$
$$\left.-\omega_b\omega_p^2\int_0^t E_z(\tau)e^{-\nu_c(t-\tau)}\cos\omega_b(t-\tau)d\tau+\frac{\partial^2 E_y}{\partial t^2}\right\},$$
$$\frac{\partial^2 E_z}{\partial x^2}=\frac{1}{c^2}\left\{\omega_p^2\int_0^t \frac{\partial E_z}{\partial \tau}e^{-\nu_c(t-\tau)}\cos\omega_b(t-\tau)d\tau+\right.$$
$$+\nu_c\omega_p^2\int_0^t E_z(\tau)e^{-\nu_c(t-\tau)}\cos\omega_b(t-\tau)d\tau+$$
$$\left.+\omega_b\omega_p^2\int_0^t E_y(\tau)e^{-\nu_c(t-\tau)}\cos\omega_b(t-\tau)d\tau+\frac{\partial^2 E_z}{\partial t^2}\right\}. \qquad \textbf{(2.45)}$$

As turned out to be obvious, applying an adequate model and a suitable coordinate system the task can be reduced to solving (2.45). As for the demanded solution is of non-monochromatic full-wave, (2.45) cannot be managed in time-domain in the common way. The exciting signal is real impulse containing starting point in time and space. The problem is transformable into complex planes in accordance with frequency and propagation-factor by using the multi-dimensional Laplace-transformation (see point 1.2).

Let the following notations be introduced for the transform-variables

$$x \longleftrightarrow p \quad \text{and} \quad t \longleftrightarrow s,$$

as in relations (1.31) and for the necessary initial conditions the relations (1.32). (As will be detailed later and was presented earlier in parts 1.3 and 2.1, it is not necessary to determine each unknown initial value separately, as there is a definite relation between these variables and the amplitude-functions. Solving the coupling equations, this relation can be obtained automatically.)

In this case, the physical meaning of (1.32) is the value of the signal entering into the magnetosphere at the boundary surface and in the starting moment. These will describe the coupling between the excitation arising in the lower atmosphere and the whistler-mode solution propagating in the magnetosphere. The plane-wave supposition affords a suitable choice of the coordinate system, which makes the elimination of some initial condition possible. In this coordinate-system the plane-polarised excitation will generate 0 initial values according to y-axis at the boundary surface, moreover – by choosing suitable $t=0$ moment – the space depending initial values according to z will be zero also

$$e_{yx0}(x)=e'_{yx0}(x)=0,\ e_{y0t}(t)=e'_{y0t}(t)=0,\ e_{zx0}(x)=e'_{zx0}(x)=0; \tag{2.46}$$

however, $e_{z0t}(t)\neq 0$ and $e'_{z0t}(t)\neq 0$.

Two initial conditions remaining will deliver the relation between the excitation and signal propagating in the plasma.

Considering all the aspects mentioned above, the transformed form of (2.45) is as follows

$$\begin{aligned}
c^2p^2E_y(p,s)&=\omega_p^2\frac{(s+\nu_c)^2}{(s+\nu_c)^2+\omega_b^2}E_y(p,s)-\\
&\quad-\omega_p^2\omega_bE_z(p,s)\frac{s+\nu_c}{(s+\nu_c)^2+\omega_b^2}+s^2E_y(p,s),\\
c^2p^2E_z(p,s)&=\omega_p^2\frac{(s+\nu_c)^2}{(s+\nu_c)^2+\omega_b^2}E_z(p,s)+\\
&\quad+\omega_p^2\omega_bE_y(p,s)\frac{s+\nu_c}{(s+\nu_c)^2+\omega_b^2}+\\
&\quad+s^2E_z(p,s)+c^2[pe_{zot}(s)+e'_{zot}(s)].
\end{aligned} \tag{2.47}$$

Introducing $e_{z0t}(s)=A(s)$ and $e'_{z0t}(s)=B(s)$ notations, the simplified form of (2.47) is

$$\begin{aligned}
E_y(p,s)&=c^2\frac{-(\omega_b\omega_p^2u)(u^2+\omega_b^2)[pA(s)+B(s)]}{[c^2p^2(u^2+\omega_b^2)-\omega_p^2u^2-s^2(u^2+\omega_b^2)]^2+[u\omega_b\omega_p^2]^2},\\
E_z(p,s)&=E_y(p,s)\frac{c^2p^2(u^2+\omega_b^2)-\omega_p^2u^2-s^2(u^2+\omega_b^2)}{-(u\omega_p^2\omega_b)},
\end{aligned} \tag{2.48}$$

where

$$u=s+\nu_c. \tag{2.48a}$$

As was expectable, the transformed forms of the electromagnetic field-components are separable and the inverse Laplace-transformation can be executed in the common way. The separated forms are

$$E_y(p,s) = c^2 \frac{a_0(s)a_\infty(s)a_2(s)}{b_3(s)} \cdot \frac{p + \dfrac{a_1(s)}{a_2(s)}}{p^4 - \dfrac{b_2(s)}{b_3(s)} p^2 + \dfrac{b_1(s)}{b_3(s)}},$$

$$E_z(p,s) = c^2 \frac{a_\infty(s)a_4(s)a_2(s)}{b_3(s)} \cdot \frac{\left[p + \dfrac{a_1(s)}{a_2(s)}\right]\left[p^2 - \dfrac{a_3(s)}{a_4(s)}\right]}{p^4 - \dfrac{b_2(s)}{b_3(s)} p^2 + \dfrac{b_1(s)}{b_3(s)}}, \qquad (2.49)$$

where

$$a_0(s) = \omega_b \omega_p^2 u, \qquad a_2(s) = A(s),$$
$$a_\infty(s) = (u^2 + \omega_b^2), \qquad a_3(s) = \omega_p^2 u^2 + s^2(u^2 + \omega_b^2),$$
$$a_1(s) = B(s), \qquad a_4(s) = (u^2 + \omega_b^2)c^2,$$
$$b_1(s) = (u\omega_b\omega_p^2)^2 + \omega_p^4 u^4 + s^4(u^2 + \omega_b^2)^2 + 2\omega_p^2 u^2 s^2 (u^2 + \omega_b^2), \qquad (2.50)$$
$$b_2(s) = 2c^2(u^2 + \omega_b^2)[\omega_p^2 u^2 + s^2(u^2 + \omega_b^2)],$$
$$b_3(s) = c^4(u^2 + \omega_b^2)^2.$$

The poles of (2.49) are

$$p_{1,2,3,4} = \pm\frac{1}{c}\sqrt{\frac{\omega_p^2(s+\nu_c)^2 + s^2[(s+\nu_c)^2 + \omega_b^2] \pm \mathrm{j}\omega_p^2\omega_b(s+\nu_c)}{(s+\nu_c)^2 + \omega_b^2}}. \qquad (2.51)$$

Detailed analysis of (2.51) is necessary for examining the influence of loss. Substituting $s = \mathrm{j}\omega$ one obtains

$$P_{1,2,3,4} = \pm\frac{1}{c}\sqrt{\frac{-\omega^2(\omega_p^2 + \omega_b^2 - \omega^2) \mp \omega_p^2\omega_b\omega + (\omega_p^2 - \omega^2)(\nu_c^2 + 2\mathrm{j}\omega\nu_c) \pm \mathrm{j}(\omega_p^2\omega_b\nu_c)}{\omega_b^2 - \omega^2 + 2\mathrm{j}\omega\nu_c + \nu_c^2}}.$$

(2.52)

Rearranging (2.52) and separating the lossy factors in the poles one can obtain, that

$$p_{1,2}=\pm\frac{1}{c}\mathrm{j}\sqrt{\frac{\omega^2\left(\omega_p^2+\omega_b^2-\omega^2\right)+\omega\omega_b\omega_p^2}{\omega_b^2-\omega^2}}\cdot\sqrt{\frac{1-\dfrac{\nu_c^2\left(\omega_b^2-\omega^2\right)+\mathrm{j}\nu_c\left[2\omega\left(\omega_p^2-\omega^2\right)+\omega_p^2\omega_b\right]}{\omega_p^2\omega_b\omega+\omega^2\left(\omega_p^2+\omega_b^2-\omega^2\right)}}{1+\dfrac{\nu_c^2+2\mathrm{j}\omega\nu_c}{\omega_b^2-\omega^2}}},$$

$$p_{3,4}=\mp\frac{1}{c}\sqrt{\frac{\left(\omega\omega_b\omega_p^2+\omega^4\right)-\omega^2\left(\omega_p^2+\omega_b^2\right)}{\omega_b^2-\omega^2}}\cdot\sqrt{\frac{1+\dfrac{\nu_c^2\left(\omega_b^2-\omega^2\right)+\mathrm{j}\nu_c\left[2\omega\left(\omega_p^2-\omega^2\right)-\omega_p^2\omega_b\right]}{\left(\omega\omega_b\omega_p^2+\omega^4\right)-\omega^2\left(\omega_p^2+\omega_b^2\right)}}{1+\dfrac{\nu_c^2+2\mathrm{j}\omega\nu_c}{\omega_b^2-\omega^2}}};$$

(2.53)

or using other notations:

$$p_{1,2}=\mp\mathrm{j}k_1(\omega)\eta_1(\omega), \qquad \textbf{(2.54a)}$$

where k_1 is the propagation factor given for a lossless electron-plasma – see relations (1.37) – and

$$\eta_1(\omega)=\sqrt[4]{\frac{1+\nu_c^2M_1(\omega)+\nu_c^4M_2(\omega)}{1+\nu_c^2N_1(\omega)+\nu_c^4N_2(\omega)}}\,\mathrm{e}^{\mathrm{j}\frac{1}{2}\left[-\mathrm{arctg}\,\nu_c\Phi_1-\mathrm{arctg}\,\nu_c\Phi_2\right]},$$

$$M_1(\omega)=\frac{\left[2\omega\left(\omega_p^2-\omega^2\right)+\omega_p^2\omega_b\right]^2-2\left(\omega_p^2-\omega^2\right)\left[\omega^2\left(\omega_p^2+\omega_b^2-\omega^2\right)+\omega_p^2\omega_b\omega\right]}{\left[\omega^2\left(\omega_p^2+\omega_b^2-\omega^2\right)+\omega_p^2\omega_b\omega\right]^2},$$

$$M_2(\omega)=\frac{\left(\omega_p^2-\omega^2\right)^2}{\left[\omega^2\left(\omega_p^2+\omega_b^2-\omega^2\right)+\omega_p^2\omega_b\omega\right]^2},$$

$$N_1(\omega)=\frac{2\left(\omega_b^2+\omega^2\right)}{\left(\omega_b^2-\omega^2\right)^2},$$

$$N_2(\omega)=\frac{1}{\left(\omega_b^2-\omega^2\right)^2}, \qquad \textbf{(2.54b)}$$

$$\Phi_1(\omega)=\frac{2\omega\left(\omega_p^2-\omega^2\right)+\omega_b\omega_p^2}{\omega^2\left(\omega_p^2+\omega_b^2-\omega^2\right)+\omega_p^2\omega_b\omega-\nu_c^2\left(\omega_p^2-\omega^2\right)},$$

$$\Phi_2(\omega)=\frac{2\omega}{\left(\omega_b^2-\omega^2\right)+\nu_c^2};$$

and further

$$p_{3,4} = \mp \alpha_3(\omega)\lambda_3(\omega)\,, \tag{2.55a}$$

where α_3 is the attenuating term of the lossless electron-plasma – see relations (1.37) – and

$$
\begin{aligned}
&\lambda_3(\omega) = \sqrt[4]{\frac{1+\nu_c^2 M'(\omega)+\nu_c^4 M_2'(\omega)}{1+\nu_c^2 N_1'(\omega)+\nu_c^4 N_2'(\omega)}}\cdot \mathrm{e}^{\mathrm{j}\frac{1}{2}[\mathrm{arctg}\,\nu_c \Phi_1' - \mathrm{arctg}\,\nu_c \Phi_2']},\\
&M_1'(\omega) = \frac{2\left(\omega_p^2-\omega^2\right)\left[\omega\omega_b\omega_p^2+\omega^4-\omega^2\left(\omega_p^2+\omega_b^2\right)\right]+\left[2\omega\left(\omega_p^2-\omega^2\right)-\omega_p^2\omega_b\right]^2}{\left[\omega\omega_b\omega_p^2+\omega^4-\omega^2\left(\omega_p^2+\omega_b^2\right)\right]^2},\\
&M_2'(\omega) = \frac{\left(\omega_p^2-\omega^2\right)^2}{\left[\omega\omega_b\omega_p^2+\omega^4-\omega^2\left(\omega_p^2+\omega_b^2\right)\right]^2},\\
&N_1'(\omega) = \frac{2\left(\omega_b^2+\omega^2\right)}{\left(\omega_b^2-\omega^2\right)^2},\\
&N_2'(\omega) = \frac{1}{\left(\omega_b^2-\omega^2\right)^2},\\
&\Phi_1'(\omega) = \frac{2\omega\left(\omega_p^2-\omega^2\right)-\omega_p^2\omega_b}{\left(\omega\omega_b\omega_p^2+\omega^4\right)-\omega^2\left(\omega_p^2+\omega_b^2\right)+\nu_c^2\left(\omega_p^2-\omega^2\right)},\\
&\Phi_2'(\omega) = \frac{2\omega}{\left(\omega_b^2-\omega^2\right)+\nu_c^2}.
\end{aligned}
\tag{2.55b}
$$

The approximate formulas in case of low loss (i.e. in case of small ν_c) are as follows

$$
\begin{aligned}
&p_{1,2} = \mp \mathrm{j}k_1(\omega)\eta_1(\omega),\\
&\eta_1(\omega) = \sqrt{1+\frac{\nu_c^2}{2}[M_1(\omega)-N_1(\omega)]}\cdot\left\{1-\mathrm{j}\frac{1}{2}\nu_c[\Phi_1(\omega)+\Phi_2(\omega)]\right\};
\end{aligned}
\tag{2.56a}
$$

and

$$p_{3,4} = \mp j\alpha_3(\omega)\lambda_3(\omega),$$

$$\lambda_3(\omega) = \sqrt{1+\frac{\nu_c^2}{2}[M_1'(\omega)-N_1'(\omega)]\cdot\left\{1+j\frac{1}{2}\nu_c[\Phi_1'(\omega)-\Phi_2'(\omega)]\right\}}. \qquad \textbf{(2.56b)}$$

The **complex form** of the lossy propagation-factors can be clearly seen in each term shown above. The meaning of this fact is well known, nevertheless covers physically essential difference in comparison to the lossless case.

The poles in case of the lossless electron-plasma are, - see relations (1.37)

$$p_1 = -jk_1, \qquad p_2 = jk_1, \qquad p_3 = -\alpha_3, \qquad p_4 = \alpha_3, \qquad \textbf{(2.57)}$$

which represent – in the $0 < \omega < \omega_b (\ll \omega_p)$ frequency range – a forward and a backward propagating mode and a "forward" and a "backward" clearly attenuated (non-propagating) mode, these latter two modes will remain at the boundary surface (the so-called "entering point") of the plasma. The mode relating to p_1 (and p_2) is called whistler-mode.

In presence of loss, this result mentioned above will be essentially modified. The attenuated modes (p_3 and p_4) will penetrate into the plasma besides the propagating **and** in this case attenuated modes (p_1 and p_2). The forms of the complex poles with the real and imaginary terms (attenuation and propagation factors) are

$$p_1 = -jk_1' - \alpha_1', \quad p_2 = jk_1' + \alpha_1', \quad p_3 = -\alpha_3' - jk_3', \quad p_4 = \alpha_3' + jk_3'. \quad \textbf{(2.58)}$$

As the plasma-model is supposed to be an infinite half space, the backward propagating (reflected) signals (p_2 and p_4) are negligible. In a plasma containing loss, two modes relating p_1 and p_3 poles will be able to propagate. The importance of this fact will be enlightened later, when the presence of ions will be examined (see points 2.3 and 2.4).

Knowing the poles shown above, the inverse Laplace-transformation of (2.49) according to p is available in the common way. This will deliver – applying the $s = j\omega$ substitution – the spectra of the electric field-components depending on x

$$E_y(x,\omega)=c^2\frac{a_0(\omega)a_\infty(\omega)a_2(\omega)}{2b_3(\omega)}\cdot\frac{1}{k_1^2(\omega)\eta_1^2(\omega)+\alpha_3^2(\omega)\lambda_3^2(\omega)}\bullet$$

$$\bullet\left\{\frac{\left[\frac{a_1(\omega)}{a_2(\omega)}-jk_1(\omega)\eta_1(\omega)\right]}{jk_1(\omega)\eta_1(\omega)}\cdot e^{-jk_1(\omega)\eta_1(\omega)x}-\right.$$

$$-\frac{\left[\frac{a_1(\omega)}{a_2(\omega)}+jk_1(\omega)\eta_1(\omega)\right]}{jk_1(\omega)\eta_1(\omega)}\cdot e^{+jk_1(\omega)\eta_1(\omega)x}-$$

$$-\frac{\left[\frac{a_1(\omega)}{a_2(\omega)}-\alpha_3(\omega)\lambda_3(\omega)\right]}{\alpha_3(\omega)\lambda_3(\omega)}\cdot e^{-\alpha_3(\omega)\lambda_3(\omega)x}+$$

$$\left.+\frac{\left[\frac{a_1(\omega)}{a_2(\omega)}+\alpha_3(\omega)\lambda_3(\omega)\right]}{\alpha_3(\omega)\lambda_3(\omega)}\cdot e^{\alpha_3(\omega)\lambda_3(\omega)x}\right\}, \tag{2.59a}$$

$$E_z(x,\omega)=c^2\frac{a_\infty(\omega)a_4(\omega)a_2(\omega)}{2b_3(\omega)}\cdot\frac{1}{k_1^2(\omega)\eta_1^2(\omega)+\alpha_3^2(\omega)\lambda_3^2(\omega)}\bullet$$

$$\bullet\left\{\frac{\left[k_1(\omega)\eta_1(\omega)+j\frac{a_1(\omega)}{a_2(\omega)}\right]\left[k_1^2(\omega)\eta_1^2(\omega)+\frac{a_3(\omega)}{a_4(\omega)}\right]}{k_1(\omega)\eta_1(\omega)}\cdot e^{-jk_1(\omega)\eta_1(\omega)x}+\right.$$

$$\left.+\frac{\left[\alpha_3(\omega)\lambda_3(\omega)-\frac{a_1(\omega)}{a_2(\omega)}\right]\left[\alpha_3^2(\omega)\lambda_3^2(\omega)-\frac{a_3(\omega)}{a_4(\omega)}\right]}{\alpha_3(\omega)\lambda_3(\omega)}\cdot e^{-\alpha_3(\omega)\lambda_3(\omega)x}\right\}. \tag{2.59b}$$

Naturally, $E_y(x,\omega)$ – demonstratively – contains terms generated by p_2 and p_4, but the reasons written earlier make them further possible to be eliminated.

The electric field components, containing only the forward propagating terms are – from (2.59a) and (2.59b), respectively

$$E_y(x,\omega)=E_{yw}(x,\omega)+E_{yN}(x,\omega),$$
$$E_z(x,\omega)=E_{zw}(x,\omega)+E_{zN}(x,\omega), \qquad \textbf{(2.59c)}$$

where the index w denotes the "whistler" mode belonging to p_1 pole and the index N denotes the "non-whistler" part of the full signal, belonging to p_3 pole.

It is important to examine and check the polarization of the propagating modes by the examination of the ratio of the field components. As is well known in the case of longitudinal propagation, a right-hand and a left-hand rotating, circularly polarised mode will appear independently in a magnetised anisotropic plasma (for monochromatic signals it is presented in e.g. *Budden*, 1966; *Ferencz*, 1996). Considering the relation of the amplitudes of E_{yw} and E_{zw} it can be found that – eliminating the identical terms in both whistler mode electric field components – the following terms remain

$$\text{in } E_y \ : \ a_0(\omega),$$
$$\text{in } E_z \ : \ a_4(\omega)\left[p_1^2(\omega)-\frac{a_3(\omega)}{a_4(\omega)}\right], \qquad \textbf{(2.60)}$$

where the $a_i(\omega)$ polynomials are defined in (2.50). Recognising that the form of the p_1 pole presented in (2.53) can be identically reformulated by $a_i(\omega)$ polynomials

$$(p_1)^2=\frac{1}{c^2}\frac{a_3(\omega)+\mathrm{j}a_0(\omega)}{a_\infty(\omega)}, \qquad \textbf{(2.61)}$$

the ratio of the two amplitudes is the following

$$E_{yw}(x,\omega)=+\mathrm{j}E_{zw}(x,\omega), \qquad \textbf{(2.62)}$$

which confirms that the whistler-mode signal is right-hand rotating – as was measured earlier.

The polarization of the attenuated modes according to p_3 will be left-hand rotating, because

$$(p_3)^2 = \frac{1}{c^2} \cdot \frac{a_3(\omega) - \mathrm{j}a_0(\omega)}{a_\infty(\omega)} \tag{2.63}$$

and

$$E_{yN}(x,\omega) = -\mathrm{j}E_{zN}(x,\omega). \tag{2.64}$$

For determining the magnetic field components, the following notations will be introduced

$$p_1 = -\mathrm{j}k_1(\omega)\eta_1(\omega) = -\mathrm{j}K_1(\omega) \text{ and } p_3 = -\alpha_3(\omega)\lambda_3(\omega) = -A_3(\omega) \tag{2.65}$$

and

$$\begin{aligned} k_1(\omega) &= \frac{\omega}{c}\sqrt{\frac{\omega_p^2}{\omega_b^2 - \omega^2}\left(\frac{\omega_b}{\omega} + 1\right) + 1} = \frac{\omega}{c}\mathcal{K}_1(\omega) = k_0(\omega) \cdot \mathcal{K}_1(\omega), \\ \alpha_3(\omega) &= \frac{\omega}{c}\sqrt{\frac{\omega_p^2}{\omega_b^2 - \omega^2}\left(\frac{\omega_b}{\omega} - 1\right) - 1} = \frac{\omega}{c}.\mathcal{A}_3(\omega) = k_0(\omega) \cdot \mathcal{A}_3(\omega). \end{aligned} \tag{2.66}$$

Further, from (2.44)

$$\begin{aligned} &H_x(x,t) = 0, \\ &H_y(x,t) = \frac{1}{\mu_0}\int^t \frac{\partial E_z(x,t)}{\partial x}\mathrm{d}t, \\ &H_z(x,t) = -\frac{1}{\mu_0}\int^t \frac{\partial E_y(x,t)}{\partial x}\mathrm{d}t \end{aligned} \tag{2.67}$$

describe the connection between $\overline{E}$ and $\overline{H}$. As is known, – see (1.40) – the Laplace transformed form of (2.67) is

$$H_y(p,s)=\frac{p}{\mu_0 s}E_z(p,s),$$

$$H_z(p,s)=-\frac{p}{\mu_0 s}E_y(p,s),$$

and – using (2.66) – it can be written, that

$$\begin{aligned}\frac{p_1}{\mu_0 s}&=-\sqrt{\frac{\varepsilon_0}{\mu_0}}\mathcal{K}_1(\omega)\eta_1(\omega)=-\frac{\mathcal{K}_1(\omega)\eta_1(\omega)}{Z_0},\\ \frac{p_3}{\mu_0 s}&=\mathrm{j}\sqrt{\frac{\varepsilon_0}{\mu_0}}\mathcal{A}_3(\omega)\lambda_3(\omega)=\mathrm{j}\frac{\mathcal{A}_3(\omega)\lambda_3(\omega)}{Z_0}.\end{aligned}\tag{2.68}$$

Applying the notations introduced above in (2.50) and in (2.65), the simplified form of E_y and E_z is as follows

$$\begin{aligned}E_y(x,\omega)&=\frac{1}{4}\left\{-\left[\frac{B(\omega)}{K_1(\omega)}-\mathrm{j}A(\omega)\right]\cdot\mathrm{e}^{-\mathrm{j}K_1(\omega)x}+\mathrm{j}\left[\frac{B(\omega)}{A_3(\omega)}-A(\omega)\right]\cdot\mathrm{e}^{-A_3(\omega)x}\right\},\\ E_z(x,\omega)&=\frac{1}{4}\left\{\mathrm{j}\left[\frac{B(\omega)}{K_1(\omega)}-\mathrm{j}A(\omega)\right]\cdot\mathrm{e}^{-\mathrm{j}K_1(\omega)x}-\left[\frac{B(\omega)}{A_3(\omega)}-A(\omega)\right]\cdot\mathrm{e}^{-A_3(\omega)x}\right\};\end{aligned}\tag{2.69}$$

further, for the magnetic field components

$$\begin{aligned}H_y(x,\omega)&=\frac{1}{4Z_0}\cdot\left\{\left[-\mathrm{j}\frac{B(\omega)}{k_0(\omega)}-\frac{K_1(\omega)A(\omega)}{k_0(\omega)}\right]\cdot\mathrm{e}^{-\mathrm{j}K_1(\omega)x}\right.\\ &\quad\left.-\mathrm{j}\left[\frac{B(\omega)}{k_0(\omega)}-\frac{A_3(\omega)A(\omega)}{k_0(\omega)}\right]\cdot\mathrm{e}^{-A_3(\omega)x}\right\},\\ H_z(x,\omega)&=\frac{1}{4Z_0}\cdot\left\{\left[-\frac{B(\omega)}{k_0(\omega)}+\mathrm{j}\frac{K_1(\omega)A(\omega)}{k_0(\omega)}\right]\cdot\mathrm{e}^{-\mathrm{j}K_1(\omega)x}\right.\\ &\quad\left.+\left[\frac{B(\omega)}{k_0(\omega)}-\frac{A_3(\omega)A(\omega)}{k_0(\omega)}\right]\cdot\mathrm{e}^{-A_3(\omega)x}\right\}.\end{aligned}\tag{2.70}$$

It is important to emphasise the meaning of $A(\omega)$ and $B(\omega)$ polynomials. These terms represent the influence of the excitation in the signal arisen in the plasma, and make the determination of the coupling at the boundary surface possible.

When all the modes generated in the plasma are known, the formal inverse Fourier-transformation may be executed and the solution containing unknown initial conditions will be obtained in time-space domain. For the propagating whistler mode this will be the following

$$
\begin{aligned}
E_i(x,t) &= \frac{1}{2\pi}\int\limits_{\omega_{min}}^{\omega_{max}} E_{i0}(x,\omega)\cdot \mathrm{e}^{\mathrm{j}[\omega t - K_1(\omega)x]}\mathrm{d}\omega,\\
H_i(x,t) &= \frac{1}{2\pi}\int\limits_{\omega_{min}}^{\omega_{max}} H_{i0}(x,\omega)\cdot \mathrm{e}^{\mathrm{j}[\omega t - K_1(\omega)x]}\mathrm{d}\omega.
\end{aligned}
\qquad \mathbf{(2.71)}
$$

The initial conditions are determinable applying the Method of Inhomogeneous Basic Modes (see Chapter 1.).

2.2.2 The excitation and the coupling

As in Fig. 1.6 can be seen, the non-monochromatic exciting current density generated by lightning discharge impulses arises in the lower atmosphere (troposphere). This medium – medium "1"– will be modelled by vacuum in the present case – as is common. The excitation in medium "1" is supposed to be a plane-wave impulse which has definite starting point according to time (t) and space (x) – see (1.45) –

$$\bar{J}_1(x,t) = J_0(x,t)\bar{e}_z\,.$$

The existing modes in half-space "1" were derived in part 1.3.4 and they are presented e.g. in equations (1.51) and (1.55), respectively.

The final task is to determine the coupling between the signals of medium "1" and "2", which is represented by the initial conditions. The Method of Inhomogeneous Basic Modes is applicable for this problem. As was discussed in details earlier (point 1.1), this method is useful in case of the sudden changes of the medium-parameters along some open or closed boundary surface(s). The separation of these media and the changes of the parameters can be described introducing distributions (i.e. the Heaviside step-function $[\Theta(x)]$ and the Dirac-delta $[\delta(x)]$ distributions). In the present case the model of the lower atmosphere is unchanged in comparison to the lossless model, so this part of the solution can be found in the previous chapter. The coupling equations to be fulfilled at the boundary surface in presence of loss are the equations (1.58). Summarising all the possibly existing modes participating in the coupling

$$
\begin{aligned}
&E_{1yR}(x,t)=\frac{1}{2\pi}\int_{-\infty}^{\infty}E_{1yR}(\omega)e^{j\omega\left(t+\frac{x}{c}\right)}d\omega,\\
&E_{1zG}(x,t)=\frac{-Z_0}{2}\int_{0}^{x_0}J_0\left(\xi,t-\frac{x-\xi}{c}\right)d\xi,\\
&E_{1zR}(x,t)=\frac{1}{2\pi}\int_{-\infty}^{\infty}E_{1zR}(\omega)e^{j\omega\left(t+\frac{x}{c}\right)}d\omega,\\
&H_{1yR}(x,t)=\frac{1}{Z_0}E_{1zR}(x,t),\\
&H_{1zR}(x,t)=-\frac{1}{Z_0}E_{1yR}(x,t),\\
&H_{1yG}(x,t)=-\frac{1}{Z_0}E_{1zG}(x,t),\\
&H_{2yw}(x,t)=-\frac{1}{8\pi}\frac{1}{Z_0}\int_{-\infty}^{\infty}\left[j\frac{B(\omega)}{k_0(\omega)}+\frac{K_1(\omega)A(\omega)}{k_0(\omega)}\right]\cdot e^{j[\omega t-K_1(\omega)(x-x_0)]}d\omega,\\
&H_{2yN}(x,t)=-\frac{1}{8\pi}\frac{j}{Z_0}\int_{-\infty}^{\infty}\left[\frac{B(\omega)}{k_0(\omega)}-\frac{A_3(\omega)A(\omega)}{k_0(\omega)}\right]\cdot e^{j\omega t-A_3(\omega)(x-x_0)}d\omega,\\
&H_{2zw}(x,t)=-jH_{2yw},\\
&H_{2zN}(x,t)=jH_{2yN},\\
&E_{2zw}(x,t)=\frac{1}{8\pi}\int_{-\infty}^{\infty}\left[j\frac{B(\omega)}{K_1(\omega)}+A(\omega)\right]\cdot e^{j[\omega t-K_1(\omega)(x-x_0)]}d\omega,\\
&E_{2zN}(x,t)=-\frac{1}{8\pi}\int_{-\infty}^{\infty}\left[\frac{B(\omega)}{A_3(\omega)}-A(\omega)\right]\cdot e^{j\omega t-A_3(\omega)(x-x_0)}d\omega,\\
&E_{2yw}(x,t)=-jE_{2zw}(x,t),\\
&E_{2yN}(x,t)=jE_{2zN}(x,t),
\end{aligned}
\tag{2.72}
$$

where the indexes are the following

R – reflected,
G – generated directly by source-current,
N – "non-whistler mode", mainly attenuated type of signals,
w – "whistler-mode" propagating type of signals,

and 1 refers to medium "1" (lower atmosphere), 2 refers to medium "2" (magnetosphere) in the indices.

With modes of (2.72) it follows that

$$
\begin{aligned}
&(1.) \quad -\frac{1}{Z_0}E_{1yR} = -\mathrm{j}H_{2yw} + \mathrm{j}H_{2yN}, \\
&(2.) \quad -\frac{1}{Z_0}E_{1zG} + \frac{1}{Z_0}E_{1zR} = H_{2yw} + H_{2yN}, \\
&(3.) \quad E_{1zG} + E_{1zR} - E_{2zw} + E_{2zN}, \\
&(4.) \quad E_{1yR} = \mathrm{j}E_{2zw} - \mathrm{j}E_{2zN},
\end{aligned}
\tag{2.73}
$$

in which the unknown variables are $A(\omega)$ and $B(\omega)$ (the initial conditions). $(1)+(4)$ in (2.73) delivers the relation between them

$$A(\omega) = \frac{-k_0(\omega)}{K_1(\omega)\cdot A_3(\omega)}B(\omega). \tag{2.74}$$

Their relation to the excitation is obtained from $(3)-(2)$ in equation (2.73): in the same way as was presented in equations (1,63)~(1.67). After all, the exact temporal-spatial function of the whistler-mode wave depending on the excitation is available. One of the field-components can be expressed as follows

$$E_{2zw}(x,t) = -\frac{Z_0}{4\pi}\int_{-\omega_{max}}^{\omega_{max}} I_{x_0}(\omega)\frac{k_0(\omega)}{k_0(\omega)+K_1(\omega)}\cdot \mathrm{e}^{\mathrm{j}[\omega t - K_1(\omega)(x-x_0)]}\mathrm{d}\omega\,, \tag{2.75}$$

where $I_{x0}(\omega)$ is given in equation (1.67) as

$$I_{x_0}(\omega) = \int_{-\infty}^{\infty}\left[\int_0^{x_0} J_0\left(\xi;t-\frac{x_0-\xi}{c}\right)\mathrm{d}\xi\right]\cdot \mathrm{e}^{-\mathrm{j}\omega t}\mathrm{d}t\,.$$

2.2.3 Solution in a weakly inhomogeneous plasma

In this point, the plasma will be considered to be weakly inhomogeneous along the superimposing magnetic field line, that is parallel to the direction of propagation (i.e. longitudinal propagation). The magnetosphere is supposed to be a one-component (electron-) plasma, which is cold, anisotropic, time-invariant and temperate, containing the effect of loss modelled by the collision of electrons.

The direction of whistler-mode propagation is $+x$ along $\bar{e}_x$ direction unit vector. As the model contains the influence of spatial inhomogeneity, the forms of the geomagnetic field and the particle- (electron-) density are

$$\overline{B}_{F0} = B_{F0}(x) \cdot \bar{e}_x \text{ and } N = N(x) \tag{2.76}$$

will be considered, further

$$\nu_c = \nu_c(x) \tag{2.77}$$

collision-frequency will describe the loss.

The plane-wave form results in

$$\frac{\partial}{\partial y} = \frac{\partial}{\partial z} = 0.$$

Maxwell's equations are given in (1.24) and the interaction equations in (2.38), however, in (2.38) $\overline{B}_F$, ν_c, and N depend on variable x. In the way presented in Chapter 1, the equations to be solved can be obtained for the unknown field components

$$E_x = 0 \quad \text{and} \quad H_x = 0,$$

$$\begin{aligned}
\frac{\partial^2 E_y}{\partial x^2} = \frac{1}{c^2}\Bigg\{&\omega_p^2(x)\int_0^t \frac{\partial E_y}{\partial \tau} e^{-\nu_c(x)(t-\tau)} \cos[\omega_b(x)\cdot(t-\tau)]d\tau + \\
&+ \nu_c(x)\omega_p^2(x)\int_0^t E_y(\tau)e^{-\nu_c(x)(t-\tau)} \cos[\omega_b(x)\cdot(t-\tau)]d\tau - \\
&- \omega_b(x)\omega_p^2(x)\int_0^t E_z(\tau)e^{-\nu_c(x)(t-\tau)} \cos[\omega_b(x)\cdot(t-\tau)]d\tau + \frac{\partial^2 E_y}{\partial t^2}\Bigg\},
\end{aligned} \tag{2.78}$$

$$\begin{aligned}
\frac{\partial^2 E_z}{\partial x^2} = \frac{1}{c^2}\Bigg\{&\omega_p^2(x)\int_0^t \frac{\partial E_z}{\partial \tau} e^{-\nu_c(x)(t-\tau)} \cos[\omega_b(x)\cdot(t-\tau)]d\tau + \\
&+ \nu_c(x)\omega_p^2(x)\int_0^t E_z(\tau)e^{-\nu_c(x)(t-\tau)} \cos[\omega_b(x)\cdot(t-\tau)]d\tau + \\
&+ \omega_b(x)\omega_p^2(x)\int_0^t E_y(\tau)e^{-\nu_c(x)(t-\tau)} \cos[\omega_b(x)\cdot(t-\tau)]d\tau + \frac{\partial^2 E_z}{\partial t^2}\Bigg\}.
\end{aligned}$$

As the two-dimensional Laplace-transformation is not applicable in this case (because the plasma-parameters depend on space), only the transformation according to time will be executed. Let the following notations be introduced

$$\int_0^\infty E_i(x,t)\cdot e^{-st}\mathrm{d}t = E_i(x,s). \tag{2.79}$$

The equations describing the field-components are as follows

$$\begin{aligned} c^2\frac{\partial^2 E_y}{\partial x^2} &= \left\{\omega_p^2(x)\frac{[s+\nu_c(x)]^2}{[s+\nu_c(x)]^2+\omega_b^2(x)}+s^2\right\}E_y(x,s)- \\ &\quad -\omega_p^2(x)\omega_b(x)\frac{[s+\nu_c(x)]}{[s+\nu_c(x)]^2+\omega_b^2(x)}E_z(x,s), \\ c^2\frac{\partial^2 E_z}{\partial x^2} &= \left\{\omega_p^2(x)\frac{[s+\nu_c(x)]^2}{[s+\nu_c(x)]^2+\omega_b^2(x)}+s^2\right\}E_z(x,s)+ \\ &\quad +\omega_p^2(x)\omega_b(x)\frac{[s+\nu_c(x)]}{[s+\nu_c(x)]^2+\omega_b^2(x)}E_y(x,s). \end{aligned} \tag{2.80}$$

For the initial conditions – similarly to homogeneous case – can be written that

$$\begin{aligned} e_{yx0}(x) &= e'_{yx0}(x) = 0, \\ e_{zx0}(x) &= e'_{zx0}(x) = 0. \end{aligned} \tag{2.81}$$

As the considered model is weakly inhomogeneous (quasi-homogeneous) – the W.K.B-method (Wentzel-Kramer-Brillouin) is applicable (see part 2.1), where the propagation factor given for homogeneous plasma forms is the basis of the inhomogeneous solution (p_i denotes the poles)

$$e^{p(s)x} \xrightarrow[(s\longleftrightarrow j\omega)]{} e^{\int_0^x p(x,s)\mathrm{d}x} \tag{2.82}$$

and the poles are

$$p_i(x,s) = \pm\frac{1}{c}\sqrt{\frac{\omega_p^2(x)[s+\nu_c(x)]^2+s^2\left[(s+\nu_c(x))^2+\omega_b^2(x)\right]\pm j\omega_p^2(x)\omega_b(x)[s+\nu_c(x)]}{[s+\nu_c(x)]^2+\omega_b^2(x)}}$$

and $i=1,2,3,4$. **(2.83)**

Though the model examined at present, is weakly inhomogeneous and because of that, the application of coupled W.K.B. is not needed when the forward and backward (reflected) propagating signals are coupled regarding the energy-balance, the possibility of this will be illustrated in the formulas. (Suitable methods for stronger inhomogeneous cases were presented in points 2.1.4 and 2.1.5.)

The poles belonging to whistler-mode signals come from (2.55a) and (2.58)

$$p_{1,2}(x,\omega)=\mp jk_1(\omega)\eta_1(\omega)=\mp[jk_1'(\omega)+\alpha_1'(\omega)]. \tag{2.84}$$

The field-components are

$$E_y(x,s)=c^2\frac{a_\infty(x,s)a_0(x,s)a_2(s)}{b_3(x,s)}\left\{\frac{\left[p_1(x,s)+\frac{a_1(s)}{a_2(s)}\right]e^{\int_0^x p_1(x,s)dx}}{[p_1(x,s)-p_2(x,s)][p_1^2(x,s)-p_3^2(x,s)]}-\right.$$

$$\left.-\frac{\left[-p_1(x,s)+\frac{a_1(s)}{a_2(s)}\right]e^{-\int_0^x p_1(x,s)dx}}{[p_1(x,s)-p_2(x,s)][p_1^2(x,s)-p_3^2(x,s)]}\right\}=E_{y1}(x,s)+E_{y2}(x,s) \tag{2.85}$$

$$E_z(x,s)=c^2\frac{a_\infty(x,s)a_4(x,s)a_2(s)}{b_3(x,s)}\left\{\frac{\left[p_1(x,s)+\frac{a_1(s)}{a_2(s)}\right]\left[p_1^2(x,s)-\frac{a_3(x,s)}{a_4(x,s)}\right]e^{\int_0^x p_1(x,s)dx}}{[p_1(x,s)-p_2(x,s)][p_1^2(x,s)-p_3^2(x,s)]}-\right.$$

$$\left.-\frac{\left[-p_1(x,s)+\frac{a_1(s)}{a_2(s)}\right]\left[p_1^2(x,s)-\frac{a_3(x,s)}{a_4(x,s)}\right]e^{-\int_0^x p_1(x,s)dx}}{[p_1(x,s)-p_2(x,s)][p_1^2(x,s)-p_3^2(x,s)]}\right\}=E_{z1}(x,s)+E_{z2}(x,s)$$

where the polynomials are

$$
\begin{aligned}
&a_0(x,s)=-\omega_b(x)\omega_p^2(x)[s+v_c(x)],\\
&a_\infty(x,s)=[s+v_c(x)]^2+\omega_b^2(x),\\
&a_1(s)=e'_{z0t}(s)=B(s),\\
&a_2(s)=e_{z0t}(s)=A(s),\\
&a_3(x,s)=\omega_p^2(x)[s+v_c(x)]^2+s^2\{[s+v_c(x)]^2+\omega_b^2(x)\},\\
&a_4(x,s)=\{[s+v_c(x)]^2+\omega_b^2(x)\}c^2=a_\infty c^2,\\
&b_1(x,s)=\{[s+v_c(x)]^2\omega_p^2(x)\omega_b(x)\}^2+\omega_p^4(x)[s+v_c(x)]^4+\\
&\qquad+s^4\{[s+v_c(x)]^2+\omega_b^2(x)\}^2+\\
&\qquad+2\omega_p^2(x)[s+v_c(x)]^2s^2\{[s+v_c(x)]^2+\omega_b^2(x)\},\\
&b_2(x,s)=2c^2\{[s+v_c(x)]^2+\omega_b^2(x)\}\cdot\{\omega_p^2(x)[s+v_c(x)]^2+\\
&\qquad+s^2\{[s+v_c(x)]^2+\omega_b^2(x)\}\},\\
&b_3(x,s)=c^4\{[s+v_c(x)]^2+\omega_b^2(x)\}^2.
\end{aligned}
\tag{2.86}
$$

As is obvious, the fact of rotation is justifiable in the way detailed earlier and

$$
E_y(x,s)=\mathrm{j}E_z(x,s)\,. \tag{2.87}
$$

After rearranging and simplifying the (2.85) and (2.86) relations, and taking into account that the lower limit of the integrals is $x=x_0$, which is the starting point of medium "2"– the following forms will be given for the field-components relating e.g. to p_1 and p_2

$$
\begin{aligned}
&E_y(x,s)=\mathrm{j}\frac{1}{4p_1(x,s)}\cdot\left\{[a_2p_1+a_1]e^{\int_{x_0}^{x}p_1(\xi,s)d\xi}-[-a_2p_1+a_1]e^{-\int_{x_0}^{x}p_1(\xi,s)d\xi}\right\},\\
&E_z(x,s)=\frac{1}{4p_1(x,s)}\cdot\left\{[a_2p_1+a_1]e^{\int_{x_0}^{x}p_1(\xi,s)d\xi}-[-a_2p_1+a_1]e^{-\int_{x_0}^{x}p_1(\xi,s)d\xi}\right\},.
\end{aligned}
\tag{2.88}
$$

and the magnetic field from Maxwell's equations has the following form

$$H_y(x,s)=\frac{1}{\mu_0 s}\frac{\partial E_z(x,s)}{\partial x},$$
$$H_z(x,s)=-\frac{1}{\mu_0 s}\frac{\partial E_y(x,s)}{\partial x}. \tag{2.89}$$

As is well known in quasi-homogeneous W.K.B. approximations, the medium parameters are slowly changing in comparison with the wavelength. This means for the monochromatic electromagnetic field that

$$\overline{G}(\overline{r},t)=\overline{G}_0(\overline{r})\cdot \mathrm{e}^{\mathrm{j}[\omega t-\varphi(\overline{r})]},$$
$$\left|\frac{\partial G_{0k}(\overline{r})}{\partial x_i}\right| \ll \left|\frac{\partial \varphi(\overline{r})}{\partial x_j}\right|. \tag{2.90}$$

In the present case the philosophy of W.K.B. is expanded for non-monochromatic signals, and for the inverse-Laplace-transformed **spectral** form of the components

$$G(\omega,x)=G_0(\omega,x)\cdot \mathrm{e}^{-\mathrm{j}k(\omega)\cdot x},$$
$$\left|\frac{\partial G_0}{\partial x}\right| \ll \left|\frac{\partial \varphi}{\partial x}\right| \tag{2.91}$$

can be supposed. Therefore (2.89) equals to the following forms

$$H_y(x,s)=\frac{p_{1,2}(x,s)}{\mu_0 s}E_z(x,s),$$
$$H_z(x,s)=-\frac{p_{1,2}(x,s)}{\mu_0 s}E_y(x,s), \tag{2.92}$$

that is

$$H_y(x,s)=\frac{1}{4\mu_0 s}\cdot\left\{[a_2 p_1+a_1]\mathrm{e}^{\int_{x_0}^{x} p_1(\xi,s)\mathrm{d}\xi}+[-a_2 p_1+a_1]\mathrm{e}^{-\int_{x_0}^{x} p_1(\xi,s)\mathrm{d}\xi}\right\},$$

$$H_z(x,s)=-\mathrm{j}\frac{1}{4\mu_0 s}\cdot\left\{[a_2 p_1+a_1]\mathrm{e}^{\int_{x_0}^{x} p_1(\xi,s)\mathrm{d}\xi}+[-a_2 p_1+a_1]\mathrm{e}^{-\int_{x_0}^{x} p_1(\xi,s)\mathrm{d}\xi}\right\}, \tag{2.93}$$

where

$$H_y(x,s) = \mathrm{j}H_z(x,s)\,. \tag{2.94}$$

In the further steps, only the forward-propagating terms of field-components will be taken into consideration, because of the reasons mentioned earlier.

Applying $s = \mathrm{j}\omega$ substitution, the p_1 pole in (2.88) is from (2.65) and (2.66)

$$p_1(\omega) = -\mathrm{j}k_1(\omega)\eta_1(\omega) = -\mathrm{j}K_1(\omega)\,. \tag{2.95}$$

As is obvious from (2.87) and (2.94) in the case of this rotating electromagnetic signal, the value of the Poynting-vector can be sufficiently examined by considering E_z and H_y

$$E_{zw}(x,\omega) = \frac{1}{4K_1(\omega)}\left[a_2(\omega)K_1(x,\omega) + \mathrm{j}a_1(\omega)\right]\mathrm{e}^{-\mathrm{j}\int_{x_0}^{x} K_1(\xi,\omega)\mathrm{d}\xi} = E_{z0}^{+}\mathrm{e}^{-\mathrm{j}\int_{x_0}^{x} K_1(\xi,\omega)\mathrm{d}\xi}\,,$$

$$H_{yw}(x,\omega) = -\frac{K_1(x,\omega)}{\mu_0\omega}E_{z0}^{+}\mathrm{e}^{-\mathrm{j}\int_{x_0}^{x} K_1(\xi,\omega)\mathrm{d}\xi} =$$

$$= -\frac{E_{z0}^{+}}{Z_0}\sqrt{\frac{\omega^2\left[\omega_p^2(x) + \omega_b^2(x) - \omega^2\right] + \omega\omega_b(x)\omega_p^2(x)}{\omega^2\left[\omega_b^2(x) - \omega^2\right]}}\cdot\eta_1(x,\omega)\mathrm{e}^{-\mathrm{j}\int_{x_0}^{x} K_1\mathrm{d}\xi}\,, \tag{2.96}$$

where Z_0 is wave impedance for vacuum. Let the following refraction-index be introduced on the basis of (E.12) for the p_1 pole belonging to the forward propagating whistler-mode signal (see Appendix E)

$$n_1(x) = \sqrt{\frac{\omega\left[\omega_p^2(x) + \omega_b^2(x) - \omega^2\right] + \omega\omega_b(x)\omega_p^2(x)}{\omega^2\left[\omega_b^2(x) - \omega^2\right]}}\cdot\eta_1(x,\omega)\,. \tag{2.97}$$

Substituting (2.97) to (2.96)

$$H_{yw}(x,\omega) = -\frac{E_{z0}^{+}}{Z_0}n_1(x)\cdot\mathrm{e}^{-\mathrm{j}\int_{x_0}^{x} K_1(\xi,\omega)\mathrm{d}\xi}\,, \tag{2.98}$$

where the index w represents the whistler-mode. The condition to be satisfied for the spectral Poynting-vector is

$$S \sim (E_{zw})(H_{yw})^{*} = (E_{z0}^{+})^{2}\frac{n_{1}(x)}{Z_{0}} = \text{const.} \triangleq (A_{0}^{+})^{2}, \tag{2.99}$$

that means the following equation

$$E_{z0}^{+} = \frac{\text{const.}}{\sqrt{n_{1}(x)}} = \frac{A_{0}^{+}}{\sqrt{n_{1}(x)}}. \tag{2.100}$$

The field-components with (2.100) are

$$\begin{aligned}
E_{zw}(x,t) &= \frac{1}{2\pi}\int_{-\infty}^{\infty} A_{0}^{+}(\omega)\sqrt[4]{\frac{\omega^{2}\left[\omega_{b}^{2}(x)-\omega^{2}\right]}{\omega^{2}\left[\omega_{p}^{2}(x)+\omega_{b}^{2}(x)-\omega^{2}\right]+\omega\omega_{b}(x)\omega_{p}^{2}(x)}}\bullet \\
&\quad \bullet\sqrt{\frac{1}{\eta_{1}(x)}}\,e^{j\left[\omega t-\int_{x_0}^{x} K_{1}(\xi,\omega)d\xi\right]}d\omega, \\
H_{yw}(x,t) &= -\frac{1}{2\pi Z_{0}}\int_{-\infty}^{\infty} A_{0}^{+}(\omega)\sqrt[4]{\frac{\omega^{2}\left[\omega_{b}^{2}(x)-\omega^{2}\right]}{\omega^{2}\left[\omega_{p}^{2}(x)+\omega_{b}^{2}(x)-\omega^{2}\right]+\omega\omega_{b}(x)\omega_{p}^{2}(x)}}\bullet \\
&\quad \bullet\sqrt{\eta_{1}(x)}\,e^{j\left[\omega t-\int_{x_0}^{x} K_{1}(\xi,\omega)d\xi\right]}d\omega,
\end{aligned} \tag{2.101}$$

where the unknown initial conditions characterising the influence of the excitation are contained in $A_{0}^{+}(\omega)$.

As a last step – similarly to the homogeneous case – the coupling equations to be solved are

$$\begin{aligned}
&(1)\quad -E_{1yR} = jZ_{0}H_{2yw} + jH_{2yN}, \\
&(2)\quad -E_{1zG} + E_{1zR} = Z_{0}H_{2yw} + H_{2yN}, \\
&(3)\quad E_{1zG} + E_{1zR} = E_{2zw} + E_{2zN}, \\
&(4)\quad E_{1yR} = jE_{2zw} - jE_{2zN},
\end{aligned} \tag{2.102}$$

where the modes arising in medium "1" are not changed. The exciting current density is identical with (1.67). From (1,102) it can be obtained that

$$A_0^+(\omega)=\frac{-\frac{Z_0}{2}I_{x_0}(\omega)}{\sqrt[4]{\frac{\delta(x_0,\omega)}{\gamma(x_0,\omega)}}\frac{1}{\sqrt{\eta_1(x_0,\omega)}}+\sqrt[4]{\frac{\gamma(x_0,\omega)}{\delta(x_0,\omega)}}\sqrt{\eta_1(x_0,\omega)}}, \tag{2.103}$$

where

$$\begin{aligned}\delta(x,\omega)&=\omega^2\left[\omega_b^2(x)-\omega^2\right],\\ \gamma(x,\omega)&=\omega^2\left[\omega_p^2(x)+\omega_b^2(x)-\omega^2\right]+\omega\omega_b(x)\omega_p^2(x).\end{aligned} \tag{2.104}$$

The whistler-mode signal in weakly inhomogeneous lossy plasma is as follows

$$\begin{aligned}E_{2zw}(x,t)&=-\frac{Z_0}{4\pi}\int_{-\omega_{max}}^{\omega_{max}}I_{x_0}(\omega)\sqrt{\frac{K_1(x_0,\omega)}{K_1(x,\omega)}}\frac{k_0(\omega)}{k_0(\omega)+K_1(x_0,\omega)}e^{j\left[\omega t-\int_{x_0}^{x}K_1(\xi,\omega)d\xi\right]}d\omega,\\ H_{2yw}(x,t)&=\frac{1}{4\pi}\int_{-\omega_{max}}^{\omega_{max}}I_{x_0}(\omega)\sqrt{\frac{K_1(x,\omega)}{K_1(x_0,\omega)}}\frac{K_1(x_0,\omega)}{k_0(\omega)+K_1(x_0,\omega)}e^{j\left[\omega t-\int_{x_0}^{x}K_1(\xi,\omega)d\xi\right]}d\omega,\end{aligned} \tag{2.105}$$

where $k_0=\omega/c$, and the form of $K_1(x,\omega)$ is known from the homogeneous solution, see (2.52), (2.53), (2.54), (2.56) and (2.97), respectively.

2.2.4 Numerical results

Further, some essential results will be presented in the case of lossy plasma. The comparison of the new results to the measurements and the former model-calculations will become valuable, if the signal processing procedure, applied for the computed signals (the solution) obtained from the new model, is identical with the one used for the former investigations. This means the application of the digital matched filtering (point 4.1.1 or *Hamar and Tarcsai,* 1982; point 4.3.1 or *Lichtenberger et al.,* 1987) and the parameter-estimating program package (FIT-program, 4.1.2 or *Tarcsai,* 1975, 1981). They have no direct connections to the presented models, except the final displaying of the results. The amplitude-time functions of the whistlers come directly from the closed-formed solution. The dynamic spectrum of the signal can be calculated from the time-function by one of the well-known and common FFT-algorithms. This fact is fundamentally new, as was not available from the former monochromatic models.

Figs. 2.16, 2.17 and 2.18 show the time-functions and the dynamic-spectra of full-wave model-whistlers for the lossless case ($\nu_c = 0$ s^{-1}) and for two different collision-frequencies ($\nu_c = 1$ s^{-1}, $\nu_c = 5$ s^{-1}) in a **homogeneous** plasma, at the spatial point x=60000km (one-hop, full path whistler ω_b =55 krad/s, ω_p =734 krad/s). The excitation is supposed to be a Dirac-delta functional. The effect of loss appears in the attenuation of the signal. This influence is not equally strong for the whole frequency-range. The lower branch of the spectrum belonging to lower frequencies can nearly disappear in case of high loss, but the amplitude-attenuation around the nose-frequency is much less.

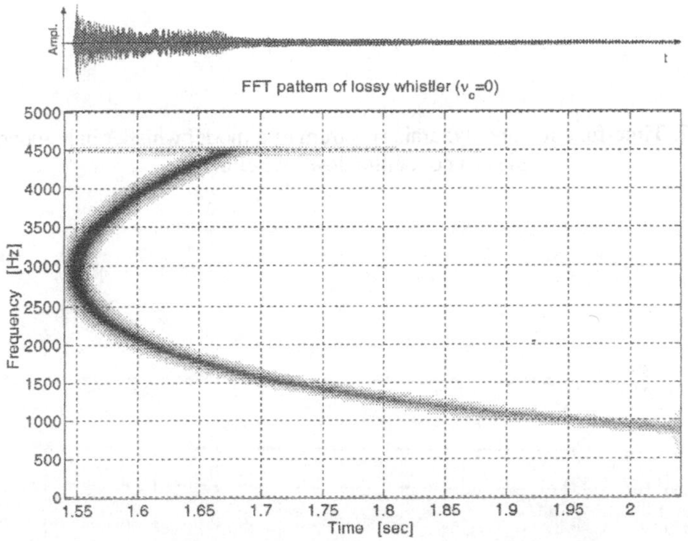

Figure 2.16. Time-function and dynamic spectrum of a model whistler in a lossless, homogeneous plasma

The time-functions and dynamic spectra of whistlers propagating in an **inhomogeneous** plasma (DE1-model, e.g. in *Tarcsai*, 1981) at L=4.654 and N_e=2550 electron/cm^3 are shown for two different collision-frequencies ($\nu_c = 0.1$ s^{-1}, $\nu_c = 5$ s^{-1}) in Figs. 2.19 and 2.20, for Dirac-delta excitation.

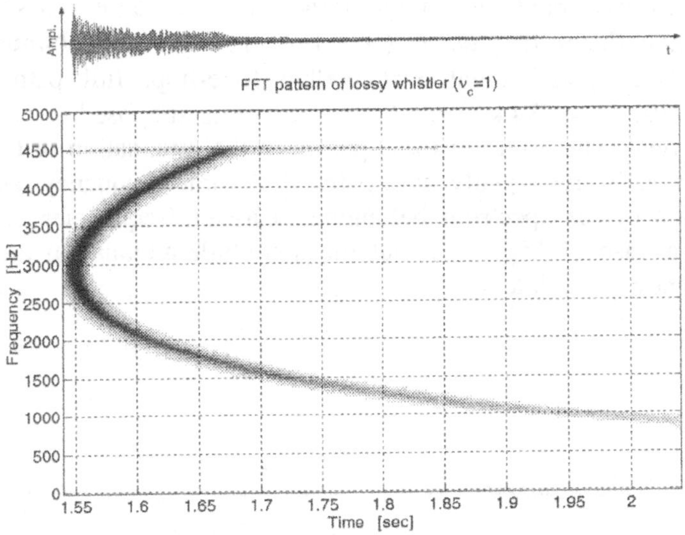

Figure 2.17. Time-function and dynamic spectrum of a model whistler in homogeneous plasma containing low loss effect

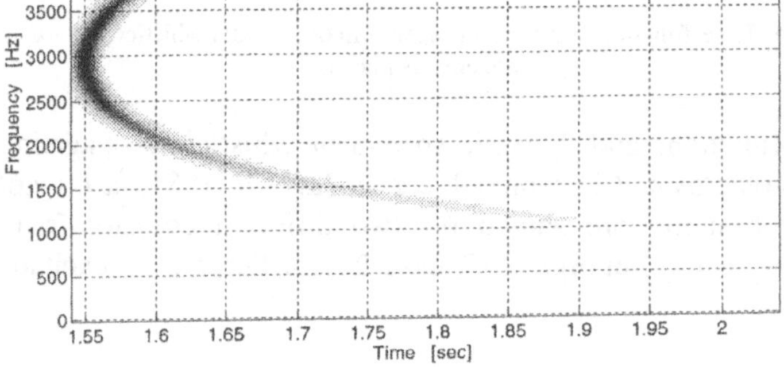

Figure 2.18. Time-function and dynamic spectrum of a model whistler in homogeneous plasma containing high loss effect

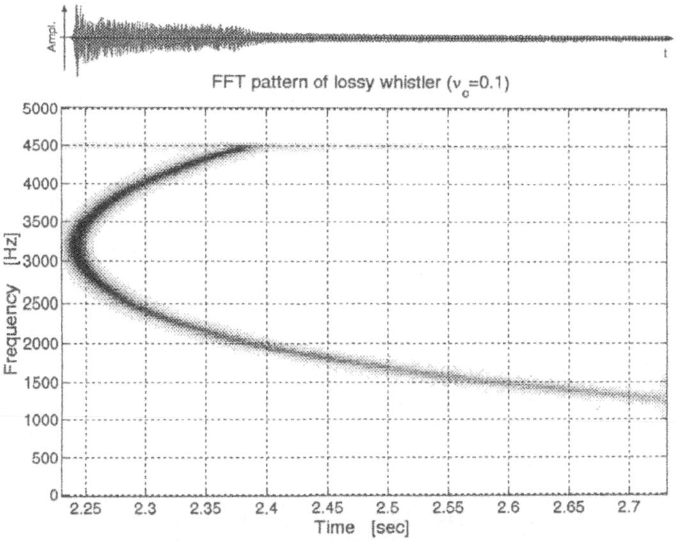

Figure 2.19. Time-function and dynamic spectrum of a model whistler in inhomogeneous plasma containing low loss effect

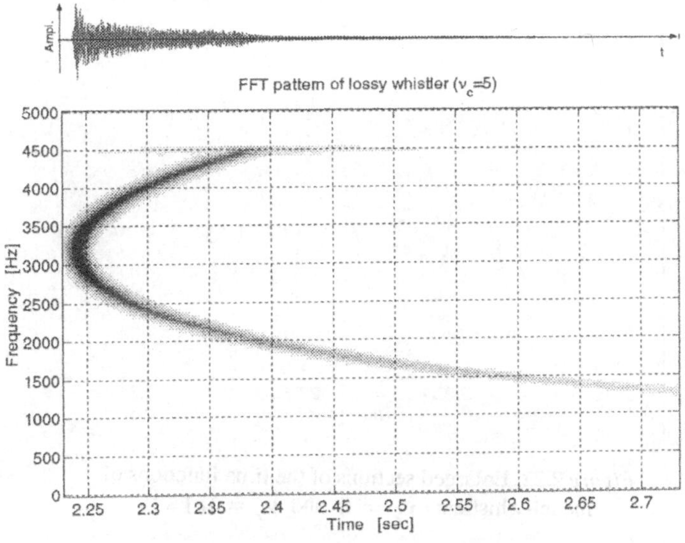

Figure 2.20. Time-function and dynamic spectrum of a model whistler in inhomogeneous plasma containing high loss effect

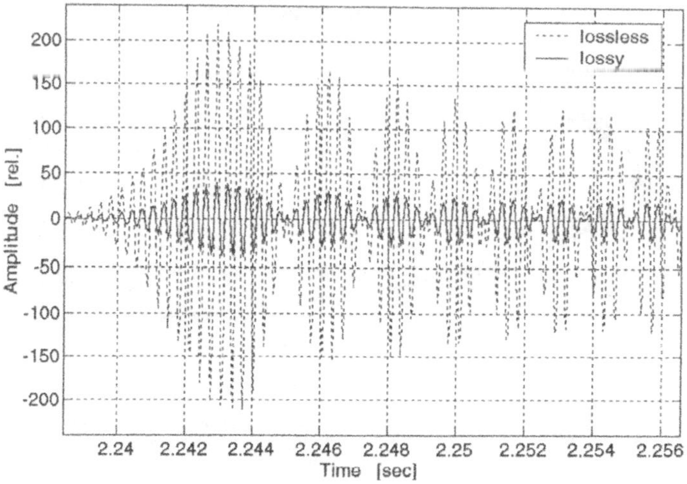

Figure 2. 21. Detailed time-functions of model-whistlers ($\nu_c = 0$ and $\nu_c = 1\ s^{-1}$)

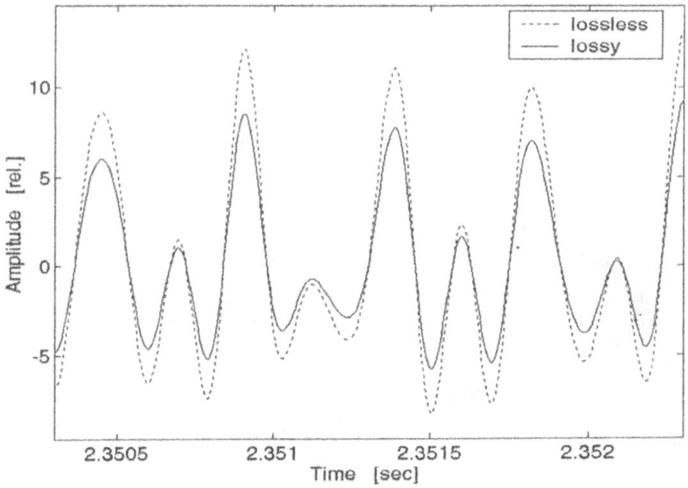

Figure 2.22. Enlarged sections of the time functions of model-whistlers ($\nu_c = 0$ and $\nu_c = 0.1\ s^{-1}$)

Enlarged, detailed patterns of whistler-mode time-functions for lossless and lossy models can be compared in Fig. 2.21 and 2.22. It is important to notice that the zero-points are not shifted in the lossy signal (the functions remain "in phase"), only the amplitude is decreased proportionately to the strength of the loss.

Fig. 2.23 shows the dynamic spectrum and the time-function of a whistler excited by a short square impulse in an inhomogeneous plasma (1.2 ms long square impulse, $\nu_c = 0.1\ s^{-1}$, L=4.654, DE1-model, the electron-density at altitude 1000km $N_e = 2550\ cm^{-3}$, x=66820). As an influence of the excitation-spectrum, minima appear periodically in the FFT-pattern of the signal. Fig. 2.24 shows a measured whistler-spectrum (Halley Bay, Antarctica). If we compare Figs. 2.23 and 2.24 it can be seen that the full-wave mathematical model shows very good coincidence with the observations, but a sensitive point of the model is the assumed form (shape) of the excitation.

The benefit of the exact full-wave description of the whistler mode propagation can obviously be seen in Fig. 2.25, where the (normalised) amplitude-frequency distributions of one measured and two calculated model-whistlers ($\nu_c = 0\ s^{-1}$ and $\nu_c = 0.2\ s^{-1}$, the supposed excitation is double square impulse, the length of the impulses $t_0 = 1.7ms$, the distance between them is $\Delta t = 0.15ms$, the ratio of their amplitudes is 1:2, the second impulse is smaller, L=4.654, x=30000km, DE1-model, $N_e = 2550\ cm^{-3}$) are shown. (The noise around the nose-frequency (*) is artefact caused by the sensitivity of the matched filtering method for the rapid changes in the signal – see point 4.1.1.) The general characters of the calculated amplitudes correlate well with the observation. The loss taken into account decreases not only the amplitude of the signal, however decreases the dynamic range of the time-dependent amplitude pattern and increases considerably the accuracy of the model making a more exact interpretation of the phenomenon possible.

2.2.5 Conclusions

One of the wave-theoretical results is the successful generalisation of the computational method for plasma models containing loss in case of arbitrarily shaped (plane-wave) excitations without common monochromatic considerations and approximations. The solution, obtained in closed form, describes the time-space dependence of the electromagnetic wave and delivers the exact amplitude-phase relations for the non-monochromatic signal, demonstrating the influence of loss on the signal. This accurate analysis of the problem was not possible by the former theories and

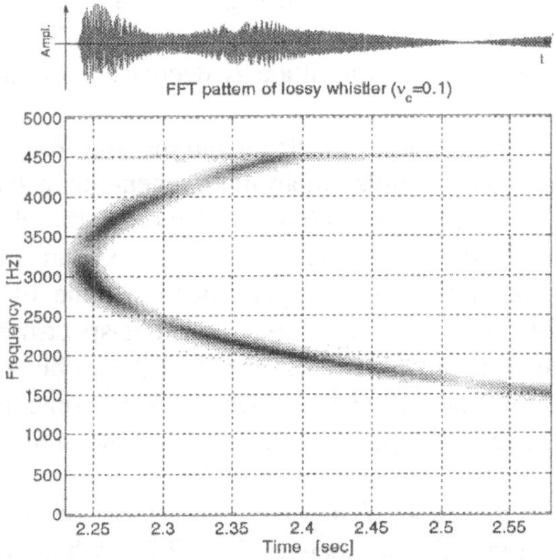

Figure 2. 23. Time-function and dynamic spectrum of whistler, excited by a short square impulse, propagating in a lossy inhomogeneous plasma

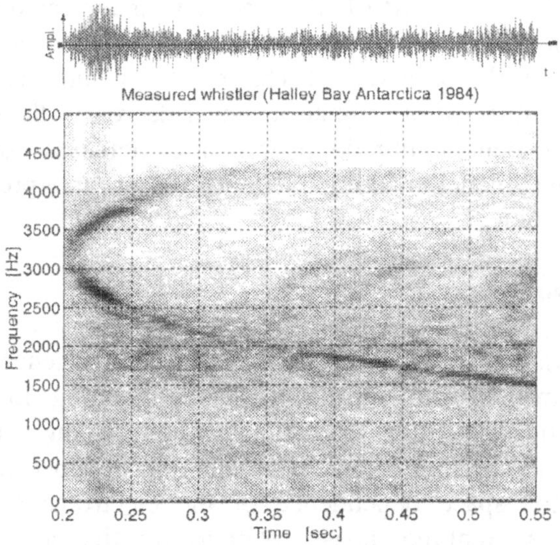

Figure 2.24. Time-function and FFT-pattern of a measured whistler (No. 84010.ns., 21. July 1984, 17:14:14 UT, Halley, Antarctica, courtesy of dr. A. J. Smith, British Antarctic Survey, Cambridge, U.K.)

approximations, because of that many hypotheses regarding the role of loss in the measured effects were born.

The influence of the loss-effect on the propagating signal is unambiguously determinable and separable from other effects.

It became theoretically justified that the presence of loss essentially changes the form of the signal, moreover, the behaviours of the arising modes are also modified. In the case of lossy electron-plasma four different modes arise (EM and HM, forward and backward directed), similarly to the lossless case, but an essential difference appears caused by the complexity of the propagation-factors. As is expected, the longitudinally propagating and right-hand rotating electron-whistler is attenuated. The left-hand rotating mode – which does not propagate in lossless plasma – penetrates into the plasma and starts to propagate (with high attenuation) only because of loss-

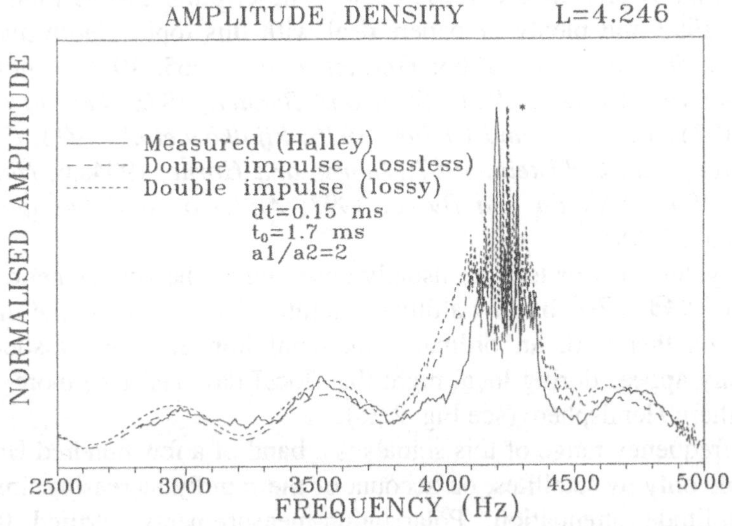

Figure 2.25. Amplitude-density functions of one measured (No. 89221AA, 9. August 1989, 16:16:56 UT, Halley, Antarctica, courtesy of dr. A. J. Smith, British Antarctic Survey, Cambridge, U.K.) and two model-whistler.

effect. (The importance of this fact will become obvious in parts 2.3 and 2.4, taking into account the ion-effects.) The propagation of this mode results in "noise"-like changes in the dynamic spectrum with a cut-off at the electron-gyro-frequency.

The application of the lossy model in whistler research makes a more accurate evaluation of the measured data possible and opens the way for new explanations of the observations.

2.3 Whistler-mode solution in a plasma containing two components (phenomenon of proton-whistlers)

2.3.1 Introduction

One of the most interesting phenomena of plasma waves is the ion-cyclotron whistler and its interaction with the plasma. As the investigation of this type of whistler waves forms a significant part of the non-monochromatic wave propagation, a brief summary of the ion-cyclotron wave-theory is presented here. A detailed description can be found e.g. in *Walker*, 1993 and plenty of papers deal with this topic (*Smith and Brice*, 1964; *Barrington et. al.*, 1966; *Gurnett et al.*, 1965, 1966; *Jones*, 1969; *Rodriguez and Gurnett*, 1971; *Terry and Budden*, 1978; *Leer et al.*,1978; *Wang*, 1971; *Watanabe and Ondoh.*, 1976; *Fijalkow et al.*, 1973; *Cserepes*, 1987; *Wong and Goldstein*, 1987; *Thorne and Horne*, 1994a,b; *Horne and Thorne*, 1994; *Villalón and Burke*, 1993; *Anderson and Fuselier*, 1994; *Singh et al.*, 1998).

Ion-cyclotron whistlers are usually observed in the geomagnetic latitude range $10° < \Theta < 70°$ by low altitude satellites. They occur in the measured spectra together with an ordinary fractional hop electron-whistler, more commonly appear during local night than local day, and also more often in the Northern Hemisphere (see Fig. 2.26).

The frequency range of this signals is a band of a few hundred Hertz and detectable only by satellites, on account of the rapidly increasing time delay and amplitude attenuation. Polarization-measurements verified that the polarization of electron- and ion-cyclotron waves is opposite (right- or left-hand elliptically or circularly polarised depending on the Northern or Southern Hemisphere and the angle between the magnetic field and the direction of propagation). Proton-, helium-, etc. whistlers are usually distinguished in the observations according to their occurrence in the pattern of dynamic spectrum.

The observation that ion-cyclotron whistlers appear to be more common in the Northern Hemisphere is thought to arise from the fact that, because of the much grater land mass, thunderstorms are much more frequent in the Northern Hemisphere than in the south.

Further the most important theories and explanations regarding the ion-cyclotron whistlers are summarised. In these approaches it is assumed that the occurrence of ion-cyclotron whistlers requires at least two ion species to occur (e.g. H^+ and He^+). On the other hand it is clearly verified deducing from Maxwell' equations (see later), that in an anisotropic plasma containing any kind of positive particles the ion-cyclotron wave may occur, so ion-whistler (proton-whistler) will propagate in a plasma containing only electrons and protons, in accordance with the measured data.

The traditional whistler-theories stand on the basis of the magneto-ionic theory (e.g. the CMA diagram of refractive index surfaces depending on a normalised frequency-scale – *Helliwell*, 1965; *Budden*, 1966; *Walker*, 1993), and consider the behaviour of a monochromatic signal in the plasma. In this theory, the longitudinal propagation seems to be an anomalous case, when no "polarization reversal" would happen. At all other angles of oblique propagation, the existence of "polarization reversal" is to be the one and only possible explanation of the ion-whistler observations.

Further, the effect of collision between the particles and the presence of spatial inhomogeneity is assumed to be indispensable in this hypothesis for the occurrence of the polarization-reversal (*Wang*, 1971; *Walker*, 1993). Moreover, it is said to be impossible for a satellite to receive an ion-whistler at any altitude for a longitudinally propagating wave in the collisionless ionosphere (*Wang*, 1971).

But, as can be seen in parts 2.2~2.5, these hypotheses are not confirmed by Maxwell's equations, in fact, the new explanation based on the real, non-monochromatic, full-wave calculations presents the continuity of the phenomenon, clearly taking into account the influence of collision, inhomogeneity, oblique propagation and ion-effects. As is detailed in the mentioned chapters, only two different forward propagating modes can arise in a linear, anisotropic plasma (EM and HM), and the influence of ions or other effects can obviously be followed in the forms of the signal. Neither the longitudinal case nor the others are anomalous, in concordance with the wave-propagation theory and with the continuous picture of the physical world. Proton-whistlers may occur in case of longitudinal propagation as well as of oblique without assuming any kind of polarisation reversal, mode-coupling or multi-component plasma. The modes are strictly the same, just their shape is modified. The fact, that the estimated direction of propagation of whistlers registered by satellites is oblique, may be originated from the same effect as the *L*-discrepancy detailed in part 2.5, however, this cannot preclude the possibility of longitudinal propagation. The presence of collision is not necessary in generation of ion-whistlers, its influence is detailed in part 2.2.

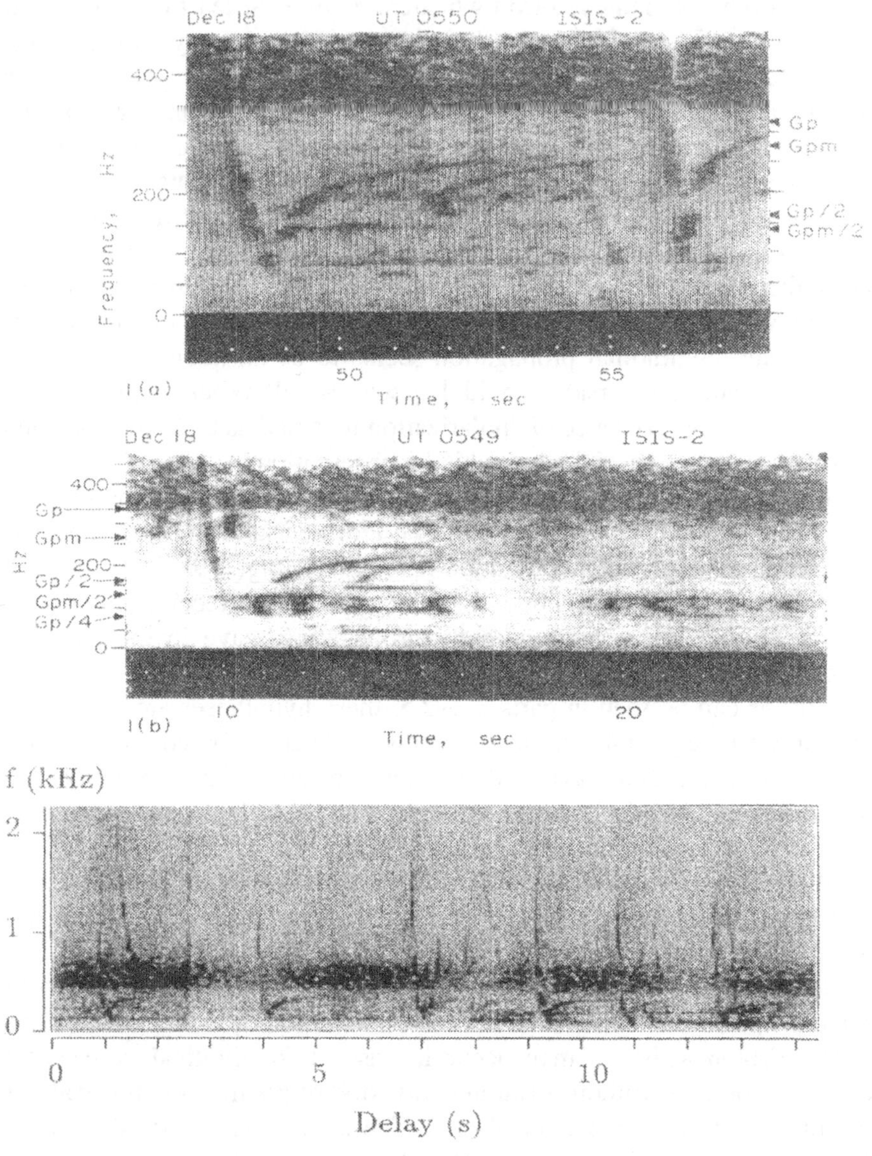

Figure 2.26 Proton-and ion-whistler registrations observed by ISIS-2 satellite in 1975 (After *Watanabe and Ondoh*, 1976, Fig. 1.; *Walker*, 1993, Fig.10.6)

A further contradiction can be found in (*Walker*, 1993), as the calculations based on the traditional theories are not verified by the observations, some parts of the model-spectra never appear in the data and some other parts of the measured signals are not contained by the model-calculations. The consistent explanation of this problem has been missed up to now. (E.g. the appearance of ion-cyclotron whistlers between different ion gyro-frequencies is unexplained by strict ray theory, and some kind of collision effect is usually taken into account in its explanation.) The theory of proton-whistlers and the model-calculations comprehend common, monochromatic approximations and result in a theory dealing with the origin of proton-whistlers, based on a crossover frequency, where a kind of energy-coupling from the electron-whistler would have to happen.

In contrast with these earlier models, the model and calculation presented here is more accurate – as for not any monochromatic approximation is contained in it – and this delivers a new explanation for the origin of the phenomenon. Since the main steps of the solving process follow that ones used for lossy plasma, only the important differences and new results will be outlined here.

The applied model:

The applied model-structure corresponds to that one shown earlier (parts 1.3, 2.1, 2.2) except only that in the present case the neutral plasma contains electrons and protons participating in the interaction between the signal and the medium, however there is no loss.

The influence of generation and recombination is ignored in the model, i.e. it is supposed that the plasma is not modified by the presence of whistler-signals. The incorporation of the effect of loss is possible in the same manner as was presented in part 2.2.

2.3.2 Solution for a homogeneous plasma

The form of Maxwell's equations to be solved is the same as in (1.24). In the plasma considered here the influence of presence of electrons and protons will be superposed. (Now the upper index "$-$" refers to electrons, while "$+$" refers to protons.) The sign of the charge of particles is included in the indices at this point.

The additional equations describing the signal-particle interactions are

$$\text{I.}\qquad m^- \frac{\partial \overline{v}^-}{\partial t} = q^-\left(\overline{E} + \overline{v}^- \times \overline{B}_{F0}\right),$$

$$\text{III.}\qquad \overline{J}^- = q^- N \overline{v}^-, \qquad \textbf{(2.106)}$$

$$\overline{\nabla} \overline{J}^- + \frac{\partial \rho^-}{\partial t} = 0,$$

and

$$\text{II.}\qquad m^+ \frac{\partial \overline{v}^+}{\partial t} = q^+\left(\overline{E} + \overline{v}^+ \times \overline{B}_{F0}\right),$$

$$\text{IV.}\qquad \overline{J}^+ = q^+ N \overline{v}^+, \qquad \textbf{(2.107)}$$

$$\overline{\nabla} \overline{J}^+ + \frac{\partial \rho^+}{\partial t} = 0.$$

Solving equations I. and II. for the velocity-components, and substituting them into equations III. and IV., the component of current-density vectors are given as

$$J_x^- = \varepsilon_0 \left(\omega_p^-\right)^2 \int^t E_x(\tau)\mathrm{d}\tau,$$

$$J_y^- = \varepsilon_0 \left(\omega_p^-\right)^2 \left\{ \frac{1}{\omega_b^-} \int^t \frac{\partial E_y(\tau)}{\partial \tau} \sin\omega_b^-(t-\tau)\mathrm{d}\tau + \int^t E_z(\tau)\sin\omega_b^-(t-\tau)\mathrm{d}\tau \right\},$$

$$J_z^- = \varepsilon_0 \left(\omega_p^-\right)^2 \left\{ \frac{1}{\omega_b^-} \int^t \frac{\partial E_z(\tau)}{\partial \tau} \sin\omega_b^-(t-\tau)\mathrm{d}\tau - \int^t E_y(\tau)\sin\omega_b^-(t-\tau)\mathrm{d}\tau \right\}, \qquad \textbf{(2.108)}$$

$$J_x^+ = \varepsilon_0 \left(\omega_p^+\right)^2 \int^t E_x(\tau)\mathrm{d}\tau,$$

$$J_y^+ = \varepsilon_0 \left(\omega_p^+\right)^2 \left\{ \frac{1}{\omega_b^+} \int^t \frac{\partial E_y(\tau)}{\partial \tau} \sin\omega_b^+(t-\tau)\mathrm{d}\tau + \int^t E_z(\tau)\sin\omega_b^+(t-\tau)\mathrm{d}\tau \right\},$$

$$J_z^+ = \varepsilon_0 \left(\omega_p^+\right)^2 \left\{ \frac{1}{\omega_b^+} \int^t \frac{\partial E_z(\tau)}{\partial \tau} \sin\omega_b^+(t-\tau)\mathrm{d}\tau - \int^t E_y(\tau)\sin\omega_b^+(t-\tau)\mathrm{d}\tau \right\},$$

where

$$\left(\omega_p^-\right)^2 = \frac{\left(q^-\right)^2 N^-}{\varepsilon_0 m^-} = \frac{e^2 N^-}{\varepsilon_0 m^-} \quad \text{in the case of electrons,}$$

$$\omega_b^- = \frac{q^- B_{F0}}{m^-} = \left(-\frac{e B_{F0}}{m^-}\right) = -\omega_{be}, \tag{2.108a}$$

$$\left(\omega_p^+\right)^2 = \frac{\left(q^+\right)^2 N^+}{\varepsilon_0 m^+} = \frac{e^2 N^+}{\varepsilon_0 m^+} \quad \text{in the case of protons,}$$

$$\omega_b^+ = \frac{q^+ B_{F0}}{m^+} = \left(+\frac{e B_{F0}}{m^+}\right),$$

$$q^- = -e,\; q^+ = +e,\; m^- \ll m^+,\; N^- = N^+ = N \quad \text{in this case.}$$

The resultant current density is

$$\bar{J} = \bar{J}^+ + \bar{J}^- . \tag{2.109}$$

Substituting these components into Maxwell's equations, $E_x \equiv H_x \equiv 0$ will be given and the differential equations-system is as follows

$$\begin{aligned}
\frac{\partial^2 E_y}{\partial x^2} = \frac{1}{c^2}\Bigg[&\left(\omega_p^+\right)^2 \int^t \frac{\partial E_y(\tau)}{\partial \tau} \cos\omega_b^+(t-\tau)\,d\tau + \\
&+\left(\omega_p^+\right)^2 \omega_b^+ \int^t E_z(\tau)\cos\omega_b^+(t-\tau)\,d\tau + \\
&+\left(\omega_p^-\right)^2 \int^t \frac{\partial E_y(\tau)}{\partial \tau} \cos\omega_b^-(t-\tau)\,d\tau + \\
&+\left(\omega_p^-\right)^2 \omega_b^- \int^t E_z(\tau)\cos\omega_b^-(t-\tau)\,d\tau + \frac{\partial^2 E_y}{\partial t^2}\Bigg], \\
\frac{\partial^2 E_z}{\partial x^2} = \frac{1}{c^2}\Bigg[&\left(\omega_p^+\right)^2 \int^t \frac{\partial E_z(\tau)}{\partial \tau} \cos\omega_b^+(t-\tau)\,d\tau - \\
&-\left(\omega_p^+\right)^2 \omega_b^+ \int^t E_y(\tau)\cos\omega_b^+(t-\tau)\,d\tau + \\
&+\left(\omega_p^-\right)^2 \int^t \frac{\partial E_z(\tau)}{\partial \tau} \cos\omega_b^-(t-\tau)\,d\tau - \\
&-\left(\omega_p^-\right)^2 \omega_b^- \int^t E_y(\tau)\cos\omega_b^-(t-\tau)\,d\tau - \frac{\partial^2 E_z}{\partial t^2}\Bigg].
\end{aligned} \tag{2.110}$$

As in lossy case, (2.110) can be Laplace-transformed according to

$x \longleftrightarrow p$ and $t \longleftrightarrow s$

variables. The initial values are

$$
\begin{aligned}
&E_i(x,t=0) = e_{ix0}(x) \stackrel{\mathcal{L}}{\longleftrightarrow} e_{ix0}(p),\\
&E_i(x=0,t) = e_{i0t}(t) \stackrel{\mathcal{L}}{\longleftrightarrow} e_{i0t}(s),\\
&\left.\frac{\partial E_i(x,t)}{\partial t}\right|_{t=0} = e_{ix0}(x) \stackrel{\mathcal{L}}{\longleftrightarrow} e'_{ix0}(p),\\
&\left.\frac{\partial E_i(x,t)}{\partial x}\right|_{x=0} = e'_{i0t}(t) \stackrel{\mathcal{L}}{\longleftrightarrow} e'_{i0t}(s).
\end{aligned}
\tag{2.111}
$$

The same considerations are valid in the present case which were detailed in earlier parts and only two independent, non-zero initial conditions will remain

$$e_{z0t}(t) \neq 0 \text{ and } e'_{z0t}(t) \neq 0. \tag{2.112}$$

For simplifying the notations, let

$$e_{z0t}(s) = A(s) \text{ and } e'_{z0t}(s) = B(s) \tag{2.113}$$

be introduced. Now the transformed form of (2.110)

$$
\begin{aligned}
c^2p^2E_y(p,s) = {} & \left(\omega_p^+\right)^2 \frac{s^2}{s^2+\left(\omega_b^+\right)^2} E_y(p,s) + \left(\omega_p^+\right)^2 \omega_b^+ \frac{s}{s^2+\left(\omega_b^+\right)^2} E_z(p,s) + \\
& + \left(\omega_p^-\right)^2 \frac{s^2}{s^2+\left(\omega_b^-\right)^2} E_y(p,s) + \left(\omega_p^-\right)^2 \omega_b^- \frac{s}{s^2+\left(\omega_b^-\right)^2} E_z(p,s) + \\
& + s^2E_y(p,s),\\
c^2p^2E_z(p,s) = {} & \left(\omega_p^+\right)^2 \frac{s^2}{s^2+\left(\omega_b^+\right)^2} E_z(p,s) - \left(\omega_p^+\right)^2 \omega_b^+ \frac{s}{s^2+\left(\omega_b^+\right)^2} E_y(p,s) + \\
& + \left(\omega_p^-\right)^2 \frac{s^2}{s^2+\left(\omega_b^-\right)^2} E_z(p,s) - \left(\omega_p^-\right)^2 \omega_b^- \frac{s}{s^2+\left(\omega_b^-\right)^2} E_y(p,s) + \\
& + s^2E_z(p,s) + c^2\left[p \cdot A(s) + B(s)\right].
\end{aligned}
\tag{2.114}
$$

As is obvious, the field-components in (2.114) are separable and their forms are

$$E_y(p,s)=c^2\frac{a_\infty(s)\cdot a_0(s)\cdot a_2(s)}{b_3(s)}\cdot\frac{p+\dfrac{a_1(s)}{a_2(s)}}{p^4-p^2\dfrac{b_2(s)}{b_3(s)}+\dfrac{b_1(s)}{b_3(s)}},$$

$$E_z(p,s)=c^2\frac{a_\infty(s)\cdot a_2(s)\cdot a_4(s)}{b_3(s)}\cdot\frac{\left[p+\dfrac{a_1(s)}{a_2(s)}\right]\left[p^2-\dfrac{a_3(s)}{a_4(s)}\right]}{p^4-p^2\dfrac{b_2(s)}{b_3(s)}+\dfrac{b_1(s)}{b_3(s)}}, \qquad \textbf{(2.115)}$$

where

$$(u^+)^2=s^2+(\omega_b^+)^2,$$
$$(u^-)^2=s^2+(\omega_b^-)^2,$$

and

$$a_\infty(s)=(u^+)^2(u^-)^2,$$
$$a_0(s)=s\left[(\omega_p^+)^2\omega_b^+(u^-)^2+(\omega_p^-)^2\omega_b^-(u^+)^2\right],$$
$$a_1(s)=B(s),$$
$$a_2(s)=A(s),$$
$$a_3(s)=s^2\left[(\omega_p^+)^2(u^-)^2+(\omega_p^-)^2(u^+)^2+(u^+)^2(u^-)^2\right],$$
$$a_4(s)=c^2(u^+)^2(u^-)^2,$$
$$b_1(s)=s^2\left[(\omega_p^+)^2\omega_b^+(u^-)^2+(\omega_p^-)^2\omega_b^-(u^+)^2\right]^2+$$
$$+s^4\left[(\omega_p^+)^2(u^-)^2+(\omega_p^-)^2(u^+)^2+(u^+)^2(u^-)^2\right]^2,$$
$$b_2(s)=2c^2(u^+)^2(u^-)^2s^2\left[(\omega_p^+)^2(u^-)^2+(\omega_p^-)^2(u^+)^2+(u^+)^2(u^-)^2\right],$$
$$b_3(s)=c^4\left[(u^-)^2\right]^2\left[(u^+)^2\right]^2.$$

The poles of (2.115) will deliver the most important information for the different behaviour of the modes arisen in the plasma.

Executing $s=j\omega$ substitution, the roots depending on frequency are

$$p_{1,2,3,4}(\omega)=\pm\frac{1}{c}\sqrt{\frac{-\omega^2\left(\omega_p^+\right)^2\mp\omega\omega_b^+\left(\omega_p^+\right)^2}{\left[\left(\omega_b^+\right)^2-\omega^2\right]}+\frac{-\omega^2\left(\omega_p^-\right)^2\mp\omega\omega_b^-\left(\omega_p^-\right)^2}{\left[\left(\omega_b^-\right)^2-\omega^2\right]}-\omega^2}\,. \quad \textbf{(2.116)}$$

Before evaluating (2.116) it is necessary to emphasise that the sign of charge of particles has still been included in the upper indices. This sign essentially influences the final form of the poles and the modes of which they belong to. As ω_b^- is negative, introducing the $|\omega_b^-|=\omega_{be}$ notation, the poles are

$$p_{1,2,3,4}(\omega)=\pm\frac{1}{c}\sqrt{\frac{-\omega^2\left(\omega_p\right)^2\mp\omega\omega_b\left(\omega_p\right)^2}{\left[\left(\omega_b\right)^2-\omega^2\right]}+\frac{-\omega^2\left(\omega_p^-\right)^2\pm\omega\omega_{be}\left(\omega_p^-\right)^2}{\left[\left(\omega_{be}\right)^2-\omega^2\right]}-\omega^2}\,, \quad \textbf{(2.117)}$$

or in separated form

$$\begin{aligned}
p_1(\omega)&=-\mathrm{j}\frac{1}{c}\sqrt{\frac{\omega^2\left(\omega_p^+\right)^2-\omega\omega_b^+\left(\omega_p^+\right)^2}{\left(\omega_b^+\right)^2-\omega^2}+\frac{\omega^2\left(\omega_p^-\right)^2+\omega\omega_{be}\left(\omega_p^-\right)^2}{\omega_{be}^2-\omega^2}+\omega^2}\,,\\
p_2(\omega)&=+\mathrm{j}\frac{1}{c}\sqrt{\frac{\omega^2\left(\omega_p^+\right)^2-\omega\omega_b^+\left(\omega_p^+\right)^2}{\left(\omega_b^+\right)^2-\omega^2}+\frac{\omega^2\left(\omega_p^-\right)^2+\omega\omega_{be}\left(\omega_p^-\right)^2}{\omega_{be}^2-\omega^2}+\omega^2}\,,\\
p_3(\omega)&=-\frac{1}{c}\sqrt{\frac{-\omega^2\left(\omega_p^+\right)^2-\omega\omega_b^+\left(\omega_p^+\right)^2}{\left(\omega_b^+\right)^2-\omega^2}-\frac{\omega^2\left(\omega_p^-\right)^2-\omega\omega_{be}\left(\omega_p^-\right)^2}{\omega_{be}^2-\omega^2}-\omega^2}\,,\\
p_4(\omega)&=+\frac{1}{c}\sqrt{\frac{-\omega^2\left(\omega_p^+\right)^2-\omega\omega_b^+\left(\omega_p^+\right)^2}{\left(\omega_b^+\right)^2-\omega^2}-\frac{\omega^2\left(\omega_p^-\right)^2-\omega\omega_{be}\left(\omega_p^-\right)^2}{\omega_{be}^2-\omega^2}-\omega^2}\,.
\end{aligned} \quad \textbf{(2.118)}$$

The structure of the poles is similar to the one presented in case of lossy plasma [see: (2.53)], but the detailed discussion of $p_{1,2,3,4}$ is needed considering that the sign of the terms may change according to the value of ω.

As is obvious, the complete $0<\omega<\omega_{be}$ frequency band is separated into two ranges, where (as is known) $\omega_b^+ \ll \omega_{be}$ (the difference is of several orders)

$$0<\omega<\omega_b^+ \text{ and } \omega_b^+<\omega<\omega_{be}\,,$$

in which the nature of poles is getting possibly changed. Further the forward directed p_1 and p_3 poles will be examined, as e.g. in part 2.2. In the pole

$$p_1 = -\mathrm{j}\frac{1}{c}\sqrt{\frac{\omega\omega_{be}\left(\omega_p^-\right)^2 + \omega^2\left[\left(\omega_p^-\right)^2 + \omega_{be}^2 - \omega^2\right]}{\omega_{be}^2 - \omega^2} - \frac{\omega\left(\omega_p^+\right)^2}{\omega_b^+ + \omega}} = -\mathrm{j}\frac{1}{c}k_1 \quad \textbf{(2.119)}$$

the term under the square-root is positive in the total $\omega < \omega_{be} (\ll \omega_p^-)$ frequency-interval, therefore the mode belonging to k_1 factor certainly propagates in the complete frequency-band. In comparison with the results for electron-plasma, it is clear that the first term in the k_1 propagation-factor is identical with the one given formerly for the so-called electron-whistler. The second term, originated from the presence of protons, modifies this propagation-factor, but does not influence its character. It can be declared, that (2.119) delivers the propagation-factor belonging to the electron-whistler. In pole

$$p_3 = -\frac{1}{c}\sqrt{\frac{-\omega^2\left(\omega_p^+\right)^2 - \omega\omega_b^+\left(\omega_p^+\right)^2}{\left(\omega_b^+\right)^2 - \omega^2} - \frac{\omega^2\left(\omega_p^-\right)^2 - \omega\omega_{be}\left(\omega_p^-\right)^2}{\omega_{be}^2 - \omega^2} - \omega^2} \quad \textbf{(2.120)}$$

the sign of the term under the square-root depends on the value of ω. While $0 < \omega < \omega_b^+$, this sign is negative, and

$$p_3 = -\mathrm{j}\frac{1}{c}\sqrt{\frac{\omega\left(\omega_p^+\right)^2}{\omega_b^+ - \omega} - \frac{\left[\omega\omega_{be}\left(\omega_p^-\right)^2 + \omega^4\right] - \omega^2\left[\left(\omega_p^-\right)^2 + \omega_{be}^2\right]}{\omega_{be}^2 - \omega^2}} = -\mathrm{j}\frac{1}{c}k_3 \quad \textbf{(2.121)}$$

can be written.

Comparing p_3 with the one given for a lossless electron-plasma – see (1.36) and (1.37) in part 1.3 –, we can recognise that the term referring to electrons, delivers the mode remaining purely attenuated at the boundary surface in the former model. In presence of protons, the first term of (2.121) will change the sign of the complete term under the square-root in the $0 < \omega < \omega_b^+$ interval, so a j can be extracted from it. This means that the mode belonging to p_3 pole will propagate in the $0 < \omega < \omega_b^+$ interval and turns into the clearly attenuated mode remaining at the boundary surface above ω_b^+ (in $\omega_b^+ < \omega < \omega_{be}$ interval). This new propagating signal arises solely because of the presence of protons in the plasma, so this results in the phenomenon of proton-whistlers.

It is important to emphasise, that the determination and discussion of these poles unambiguously verified the fact that the electron- and proton-whistlers come into being and propagate completely independently from each other (and from any hypothetical cross-over frequency, e.g. *Walker*,

1993). Consequently, there is no need for any frequency-dependent hypothesis to explain the origin of proton-whistlers, for example the well-known polarization-reversal of electron-whistlers. The proton-whistlers principally cannot arise from the electron-whistlers by the aid of any energy coupling-out, because these parts of the propagating electro-magnetic field exist independently. This fact is independent of the longitudinal or oblique directed propagation (see part 2.5).

These results coincide completely with the traditional electro-magnetic field theory concerning plasmas (cf. Faraday-rotation).

Moreover, as a further consequence of that result, it is obvious in comparison with the effect of loss, that the presence of ions or the existence of loss in the plasma will result in substantially the same influence, because the same part of the signal will change its nature from a clearly attenuated mode to a propagating one.

The examination of rotating polarization for the two forward propagating signal can be executed in the way detailed in part. 2.2. The mode given for electron-whistlers is right-handed rotating, while the mode belonging to p_3 pole is left-handed rotating, in concordance with the measured data.

Applying the mathematical steps used earlier (see e.g. part. 2.2) the magnetic field components can be written as

$$H_y(x,t)=\frac{1}{\mu_0}\int^t \frac{\partial E_z(x,t)}{\partial x}dt,$$
$$H_z(x,t)=-\frac{1}{\mu_0}\int^t \frac{\partial E_y(x,t)}{\partial x}dt, \tag{2.122}$$

and the transformed forms are

$$H_y(p,s)=\frac{p}{\mu_0 s}E_z(p,s),$$
$$H_z(s,p)=-\frac{p}{\mu_0 s}E_y(p,s). \tag{2.123}$$

The spectral field-components are (in $0<\omega<\omega_b^+$ range)

$$E_{2y}(x,\omega)=-\frac{1}{4}\left\{\left(\frac{B(\omega)}{k_1}-jA(\omega)\right)\cdot e^{-jk_1(x-x_0)}-\right.$$

$$\left.-\left(\frac{B(\omega)}{k_3}-jA(\omega)\right)\cdot e^{-jk_3(x-x_0)}\right\},$$

$$E_{2z}(x,\omega)=\frac{1}{4}\left\{j\left(\frac{B(\omega)}{k_1}-jA(\omega)\right)\cdot e^{-jk_1(x-x_0)}+\right.$$

$$\left.+j\left(\frac{B(\omega)}{k_3}-jA(\omega)\right)\cdot e^{-jk_3(x-x_0)}\right\},$$

$$H_{2y}(x,\omega)=\frac{1}{4Z_0}\left\{\left(-j\frac{B(\omega)}{k_0}-\frac{k_1A(\omega)}{k_0}\right)\cdot e^{-jk_1(x-x_0)}-\right.$$

$$\left.-j\left(\frac{B(\omega)}{k_0}-\frac{jk_3A(\omega)}{k_0}\right)\cdot e^{-jk_3(x-x_0)}\right\},$$

$$H_{2z}(x,\omega)=-\frac{1}{4Z_0}\left\{\left(\frac{B(\omega)}{k_0}-\frac{jk_1A(\omega)}{k_0}\right)\cdot e^{-jk_1(x-x_0)}-\right.$$

$$\left.-\left(\frac{B(\omega)}{k_0}-\frac{jk_3A(\omega)}{k_0}\right)\cdot e^{-jk_3(x-x_0)}\right\}; \qquad \textbf{(2.124a)}$$

and in the $\omega_b^+ < \omega < \omega_{be}$ range:

$$E_{2y}(x,\omega)=-\frac{1}{4}\left\{\left(\frac{B(\omega)}{k_1}-jA(\omega)\right)\cdot e^{-jk_1(x-x_0)}\right.$$

$$\left.-j\left(\frac{B(\omega)}{\alpha_3}-A(\omega)\right)\cdot e^{-\alpha_3(x-x_0)}\right\}, \qquad \textbf{(2.124b)}$$

$$E_{2z}(x,\omega)=\frac{1}{4}\left\{j\left(\frac{B(\omega)}{k_1}-jA(\omega)\right)\cdot e^{-jk_1(x-x_0)}-\right.$$

$$\left.-\left(\frac{B(\omega)}{\alpha_3}-A(\omega)\right)\cdot e^{-\alpha_3(x-x_0)}\right\}, \qquad \text{cont.}$$

$$H_{2y}(x,\omega)=\frac{1}{4Z_0}\left\{\left(-\mathrm{j}\frac{B(\omega)}{k_0}-\frac{k_1A(\omega)}{\mathrm{k}_0}\right)\cdot \mathrm{e}^{-\mathrm{j}k_1(x-x_0)}-\right.$$
$$\left.-\mathrm{j}\left(\frac{B(\omega)}{k_0}-\frac{\alpha_3A(\omega)}{k_0}\right)\cdot \mathrm{e}^{-\alpha_3(x-x_0)}\right\},$$
$$H_{2z}(x,\omega)=-\frac{1}{4Z_0}\left\{\left(\frac{B(\omega)}{k_0}-\frac{\mathrm{j}k_1A(\omega)}{\mathrm{k}_0}\right)\cdot \mathrm{e}^{-\mathrm{j}k_1(x-x_0)}-\right.$$
$$\left.-\left(\frac{B(\omega)}{k_0}-\frac{\alpha_3A(\omega)}{k_0}\right)\cdot \mathrm{e}^{-\alpha_3(x-x_0)}\right\}.$$
(2.124b)

Introducing notations in the indices (*ew* – electron-whistlers; *pw* – proton-whistlers; N – non-propagating)

$$\begin{aligned} &0<\omega<\omega_b^+, && \omega_b^+<\omega<\omega_{be}, \\ &E_{2y}=E_{2yew}+E_{2ypw}, && E_{2y}=E_{2yew}+E_{2yN}, \\ &E_{2z}=E_{2zew}+E_{2zpw}, && E_{2z}=E_{2zew}+E_{2zN}, \\ &H_{2y}=H_{2yew}+H_{2ypw}, && H_{2y}=H_{2yew}+H_{2yN}, \\ &H_{2z}=H_{2zew}+H_{2zpw}, && H_{2z}=H_{2zew}+H_{2zN}. \end{aligned} \qquad \textbf{(2.125)}$$

As in medium "1" there is no modification, the results presented in Chapter 1 are valid for the modes arisen

$$\begin{aligned} &E_{1yR}(x,t)=\frac{1}{2\pi}\int_{-\infty}^{\infty}E_{1yR}(\omega)\mathrm{e}^{\mathrm{j}\omega\left(t+\frac{x}{c}\right)}\mathrm{d}\omega, && H_{1yR}(x,t)=\frac{1}{Z_0}E_{1zR}(x,t), \\ &E_{1zG}(x,t)=\frac{-Z_0}{2}\int_0^{x_0}J_0\left(\xi,t-\frac{x-\xi}{c}\right)\mathrm{d}\xi, && H_{1zR}(x,t)=-\frac{1}{Z_0}E_{1yR}(x,t), \\ &E_{1zR}(x,t)=\frac{1}{2\pi}\int_{-\infty}^{\infty}E_{1zR}(\omega)\mathrm{e}^{\mathrm{j}\omega\left(t+\frac{x}{c}\right)}\mathrm{d}\omega, && H_{1yG}(x,t)=-\frac{1}{Z_0}E_{1zG}(x,t). \end{aligned} \qquad \textbf{(2.126)}$$

The coupling equations are to be solved in both ranges

$$0<\omega<\omega_b^+,$$

$$(1)\quad -\frac{1}{Z_0}E_{1yR}=-\mathrm{j}H_{2yew}+\mathrm{j}H_{2ypw},$$

$$(2)\quad -\frac{1}{Z_0}E_{1zG}+\frac{1}{Z_0}E_{1zR}=H_{2yew}+H_{2ypw},$$

$$(3)\quad E_{1zG}+E_{1zR}=E_{2zew}+E_{2zpw},$$

$$(4)\quad E_{1yR}=\mathrm{j}E_{2zew}-\mathrm{j}E_{2zpw};$$

(2.127)

$$\omega_b^+<\omega<\omega_{be},$$

$$(1)\quad -\frac{1}{Z_0}E_{1yR}=-\mathrm{j}H_{2yew}+\mathrm{j}H_{2yN},$$

$$(2)\quad -\frac{1}{Z_0}E_{1zG}+\frac{1}{Z_0}E_{1zR}=H_{2yew}+H_{2yN},$$

$$(3)\quad E_{1zG}+E_{1zR}=E_{2zew}+E_{2zN},$$

$$(4)\quad E_{1yR}=\mathrm{j}E_{2zew}-\mathrm{j}E_{2zN}.$$

Applying (1.67) the final forms of the electromagnetic field in region "2" propagating in electron and proton whistler-mode (for one field component, if $0<\omega<\omega_b^+$) are as

$$E_{2zew}=-\frac{Z_0}{4\pi}\int_{-\omega_{max}}^{\omega_{max}}I_{x0}(\omega)\frac{k_0(\omega)}{k_0(\omega)+k_1(\omega)}\mathrm{e}^{\mathrm{j}[\omega t-k_1(\omega)(x-x_0)]}\,\mathrm{d}\omega,$$

$$E_{2zpw}=-\frac{Z_0}{4\pi}\int_{-\omega_{max}}^{\omega_{max}}I_{x0}(\omega)\frac{k_0(\omega)}{k_0(\omega)+k_3(\omega)}\mathrm{e}^{\mathrm{j}[\omega t-k_3(\omega)(x-x_0)]}\,\mathrm{d}\omega;$$

(2.128)

and in the $\omega_b^+<\omega<\omega_{be}$ range

$$E_{2zew}=-\frac{Z_0}{4\pi}\int_{-\omega_{max}}^{\omega_{max}}I_{x0}(\omega)\frac{k_0(\omega)}{k_0(\omega)+k_1(\omega)}\mathrm{e}^{\mathrm{j}[\omega t-k_1(\omega)(x-x_0)]}\,\mathrm{d}\omega,$$

$$E_{2zN}=\frac{Z_0}{4\pi}\int_{-\omega_{max}}^{\omega_{max}}I_{x0}(\omega)\frac{k_0(\omega)}{\mathrm{j}\alpha_3(\omega)-k_0(\omega)}\mathrm{e}^{\mathrm{j}\omega t-\alpha_3(\omega)(x-x_0)}\,\mathrm{d}\omega.$$

(2.129)

2.3.3 Solution for a weakly inhomogeneous plasma

The spatial inhomogeneity in the plasma will be taken into consideration by

$$\omega_h^+(x),\ \omega_p^+(x),\ \omega_h^-(x),\ \omega_p^-(x),\ N(x),\ \overline{B}_{F0}(x)$$

space-dependent values of plasma-parameters (and the superimposing magnetic field) – as was mentioned earlier in Chapter 2.

The differential equation-system in time-space domain is as follows

$$\frac{\partial^2 E_y}{\partial x^2}=\frac{1}{c^2}\Bigg\{\frac{[\omega_p^+(x)]^2}{\omega_b^+(x)}\int^t\frac{\partial E_y}{\partial\tau}\omega_b^+(x)\cdot\cos\omega_b^+(x)(t-\tau)\mathrm{d}\tau+$$

$$+[\omega_p^+(x)]^2\int^t E_z(\tau)\cdot\omega_b^+(x)\cdot\cos\omega_b^+(x)(t-\tau)\mathrm{d}\tau+$$

$$+\Bigg\{\frac{[\omega_p^-(x)]^2}{\omega_b^-(x)}\int^t\frac{\partial E_y}{\partial\tau}\omega_b^-(x)\cdot\cos\omega_b^-(x)(t-\tau)\mathrm{d}\tau+$$

$$+[\omega_p^-(x)]^2\int^t E_z(\tau)\cdot\omega_b^-(x)\cdot\cos\omega_b^-(x)(t-\tau)\mathrm{d}\tau+\frac{\partial^2 E_y}{\partial t^2}\Bigg\},$$

$$\frac{\partial^2 E_z}{\partial x^2}=\frac{1}{c^2}\Bigg\{\frac{[\omega_p^+(x)]^2}{\omega_b^+(x)}\int^t\frac{\partial E_z}{\partial\tau}\omega_b^+(x)\cdot\cos\omega_b^+(x)(t-\tau)\mathrm{d}\tau-$$

$$-[\omega_p^+(x)]^2\int^t E_y(\tau)\cdot\omega_b^+(x)\cdot\cos\omega_b^+(x)(t-\tau)\mathrm{d}\tau+$$

$$+\Bigg\{\frac{[\omega_p^-(x)]^2}{\omega_b^-(x)}\int^t\frac{\partial E_z}{\partial\tau}\omega_b^-(x)\cdot\cos\omega_b^-(x)(t-\tau)\mathrm{d}\tau-$$

$$-[\omega_p^-(x)]^2\int^t E_y(\tau)\cdot\omega_b^-(x)\cdot\cos\omega_b^-(x)(t-\tau)\mathrm{d}\tau+\frac{\partial^2 E_z}{\partial t^2}\Bigg\}. \qquad \textbf{(2.130)}$$

Applying the computational method outlined in earlier parts of this chapter the solution can be determined using the W.K.B. philosophy, in which the propagation factor given for homogeneous plasma will be transferred into the formulas

$$\mathrm{e}^{p_i(s)\cdot x}\longrightarrow\mathrm{e}^{\int_0^x p_i(x,s)\mathrm{d}x}, \qquad \textbf{(2.131)}$$

and in first approximation the Poynting-vector is assumed to be constant for the forward and the backward propagating modes independently $S \sim n(x) \cdot E_0^2 = \text{const.}$, as was applied earlier.

The accurate form of the solution in a weakly inhomogeneous plasma containing electrons and protons is the following (*ew* denotes the electron-, *pw* denotes the proton-whistlers in the indices)

$$E_{2zew}(x,t) = -\frac{Z_0}{4\pi} \int_{-\omega_{max}}^{\omega_{max}} I_{x0}(\omega) \sqrt{\frac{k_1(x_0,\omega)}{k_1(x,\omega)}} \frac{k_0(\omega)}{k_0(\omega)+k_1(x_0,\omega)} \cdot e^{j\left[\omega t - \int_{x0}^{x} k_1(\xi,\omega)d\xi\right]} d\omega ,$$

$$H_{2yew}(x,t) = \frac{1}{4\pi} \int_{-\omega_{max}}^{\omega_{max}} I_{x0}(\omega) \sqrt{\frac{k_1(x,\omega)}{k_1(x_0,\omega)}} \frac{k_1(x_0,\omega)}{k_0(\omega)+k_1(x_0,\omega)} \cdot e^{j\left[\omega t - \int_{x0}^{x} k_1(\xi,\omega)d\xi\right]} d\omega ,$$

$$E_{2zpw}(x,t) = -\frac{Z_0}{4\pi} \int_{-\omega_{max}}^{\omega_{max}} I_{x0}(\omega) \sqrt{\frac{k_3(x_0,\omega)}{k_3(x,\omega)}} \frac{k_0(\omega)}{k_0(\omega)+k_3(x_0,\omega)} \cdot e^{j\left[\omega t - \int_{x0}^{x} k_3(\xi,\omega)d\xi\right]} d\omega ,$$

$$H_{2ypw}(x,t) = \frac{1}{4\pi} \int_{-\omega_{max}}^{\omega_{max}} I_{x0}(\omega) \sqrt{\frac{k_3(x,\omega)}{k_3(x_0,\omega)}} \frac{k_3(x_0,\omega)}{k_0(\omega)+k_3(x_0,\omega)} \cdot e^{j\left[\omega t - \int_{x0}^{x} k_3(\xi,\omega)d\xi\right]} d\omega ,$$

(2.132)

where $k_0 = \omega / c$ and $I_{x0}(\omega) = \int_{-\infty}^{\infty} \left[\int_{0}^{x_0} J_0\left(\xi, t - \frac{x_0 - \xi}{t}\right) d\xi \right] \cdot e^{-j\omega t} dt$.

2.3.4 Numerical results

The exact formulas presented in part 2.3.3 were built into the computer-codes. Some important results are presented in this part. As the most interesting features well can be seen in the case of Dirac-delta excitation, the more complicated excitation-functions are not dealt with in this chapter, because the transfer function characterises the propagation itself.

As the former approaches based on the monochromatic magneto-ionic theory have no description – even a rough one – regarding the time-space function and the shape of the signal and they contain only hypotheses concerning the generation and propagation mechanism of the proton-whistlers, the presented results are fundamentally new and the way is opened to apply the signal processing and analysing methods common for electron-whistlers in the case of proton-whistlers also, giving the closed formed time-space function of the signal. Because of the reasons mentioned above, no matched filtering process has been developed up to now for ion-whistlers, and only the FFT process is available for effectual comparison of the simulated and measured data.

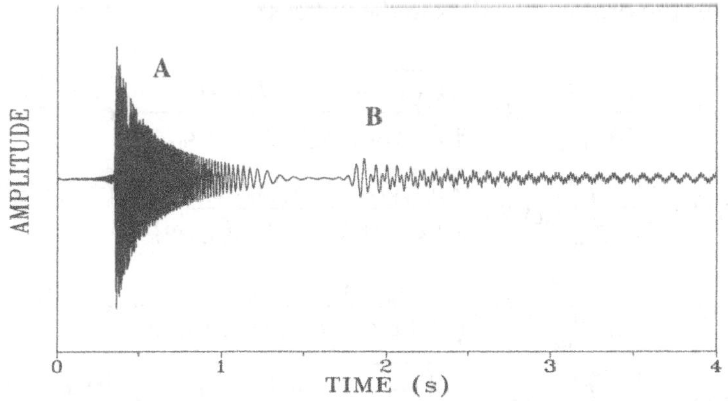

Figure 2.27 .Time-function of a complete (electron- and proton-) whistler in a homogeneous plasma

The time-function of the proton- and the electron-whistler propagating in a homogeneous plasma is shown in Fig. 2.27 (x=7000km). The first part of the signal is the electron-whistler (denoted by A), however, in the second part of the wave-pattern (from ~ 1.7s) the proton-whistler becomes dominant (denoted by B).

In Figs. 2.28, 2.29, 2.30 and 2.31 the dynamic spectra of the electron- and proton-whistlers can be seen at a given height (x=7000km), considering different proton-electron ratios (r=*85% - 100%*) in a homogeneous, neutral plasma (as an approximate model of the influence of heavier positive ions not participating in the signal-medium interaction, $N_e = 1000\ cm^{-3}$). The first, long, steep line with high amplitude is the electron-whistler, the slower mode is the proton-whistler in the spectrum. It is clearly shown that the spectrum is very sensitively influenced by the ratio of the interactive particles. In pure proton-electron plasma the lower branch of the proton-whistler disappears. In the case of proton-electron ratios lower than 100%, the lower segment of the proton-whistler becomes more visible and the distance between the electron- and proton-whistler increases.

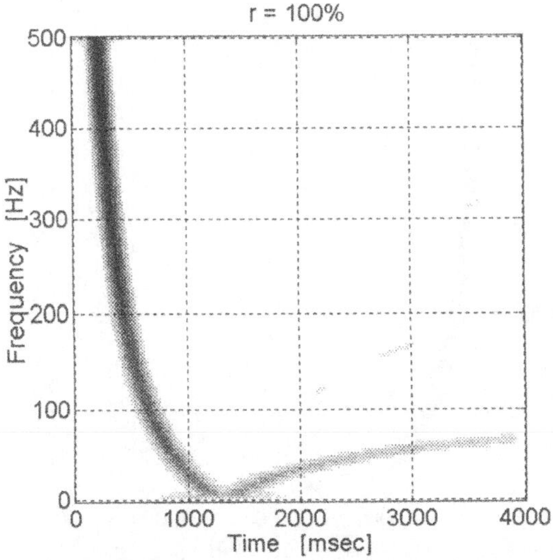

Figure 2.28. FFT pattern of an electron-proton whistler pair in a homogeneous plasma, the electron-proton ratio r=100%

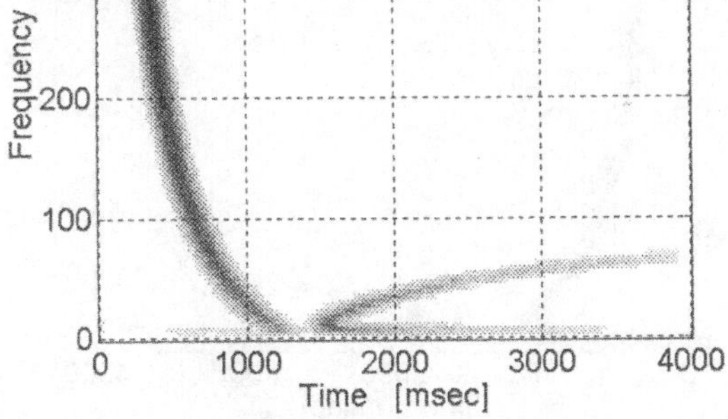

Figure 2.29. FFT pattern of an electron-proton whistler pair in a homogeneous plasma, the electron-proton ratio r=95%

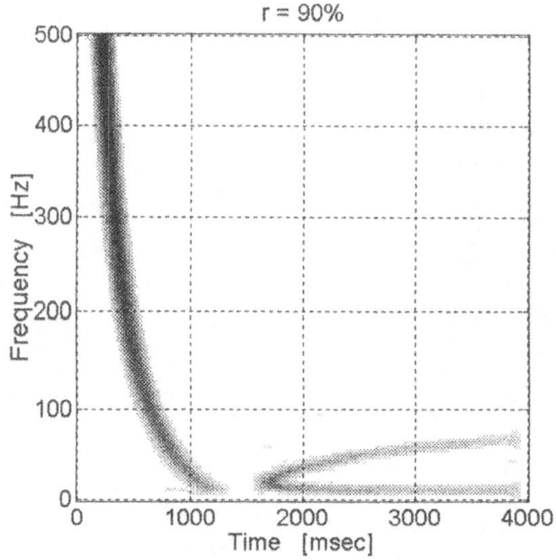

Figure 2.30. FFT pattern of an electron-proton whistler pair in a homogeneous plasma, the electron-proton ratio r=90%

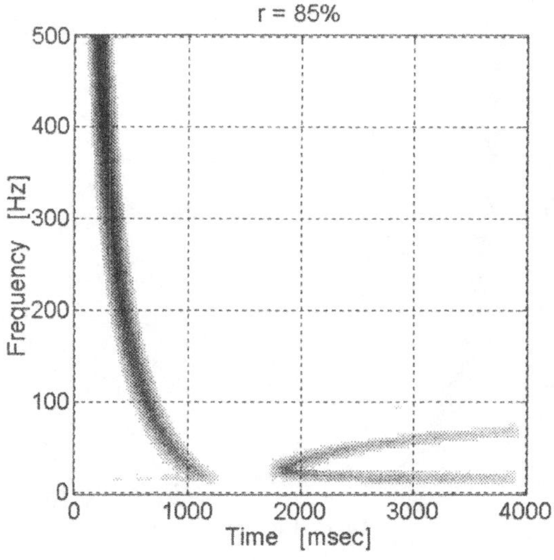

Figure 2.31. FFT pattern of an electron-proton whistler pair in a homogeneous plasma, the electron-proton ratio r=85%

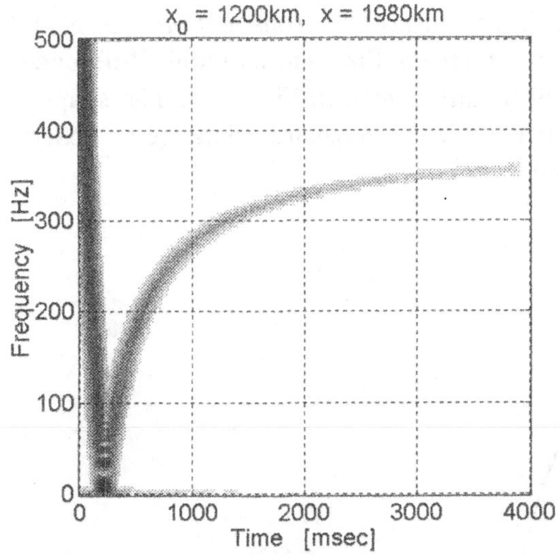

Figure 2.32. FFT-pattern of an electron-proton whistler-pair in inhomogeneous plasma, $r = 100\%$

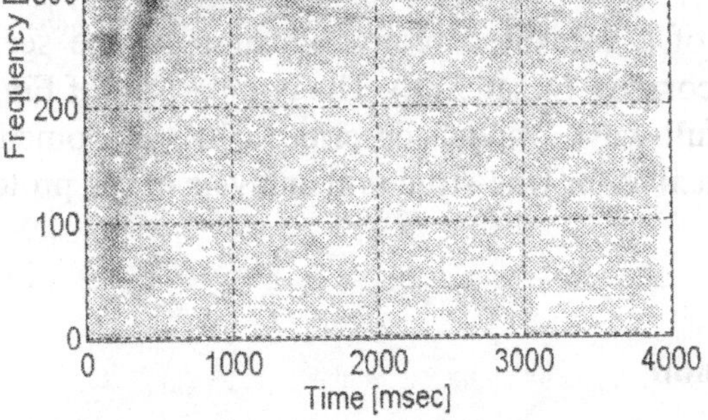

Figure 2.33. Measured proton-whistler (Intercosmos-14 satellite)

Fig. 2.32 shows the FFT-pattern of a proton-whistler in an inhomogeneous plasma-model, the signal penetrates into the plasma at x_0 and propagates to x, the ratio of electrons and protons is 100%. The electron-density function used (from the **I**nternational **R**eference **I**onosphere, IRI model, *Bilitza*, 1990) can be seen in Fig. 2.34. The shape of the spectrum is in good coincidence with the measured data, (e.g. observations of ISIS-2, 1975, see Fig 2.26).

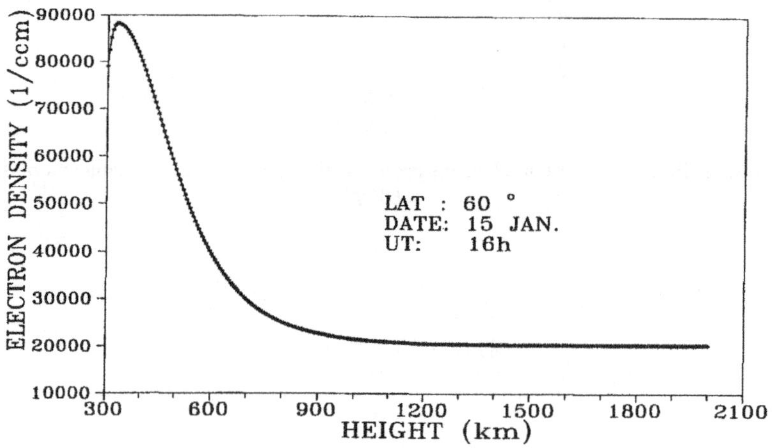

Figure 2.34. Electron-density distribution function based on the IRI model

The spectrum of a measured proton-whistler can be seen in Fig. 2.33 (recorded by Intercosmos-14 satellite). The comparison of Fig. 2.32 and 2.33 shows that the full-wave model describes the phenomenon with high accuracy and applicable for fine-structural analysis of the proton-whistlers.

2.3.5 Conclusion

It is possible to extend the application of the theoretical model and the computational method for magnetised plasma containing ion-effects, in the case of arbitrary plane wave (e.g. impulse-type) excitation. The model does not contain monochromatic considerations (e.g. wave-packet, harmonicwaves, Appleton-Hartree formula for refractive indices, dielectric tensor, etc). The new theoretical model makes the development of the analytical processes (e.g. digital matched filtering) possible for proton (ion-) whistlers, and has opened the way for the fine-structural analysis of the observations.

As was expected based upon the continuity of the wave-propagation theory, the four modes arise again (forward and backward directed EM and HM) which appear also in (lossless or lossy) electron-plasma. The right-hand rotating (HM) mode propagates up to the electron-gyro-frequency and this is the electron-whistler. The left-hand rotating (EM) mode propagates up to the proton-gyro-frequency and this is the proton-whistler. This mode turns into a clearly attenuated signal remaining at the plasma-boundary surface above the proton-gyro-frequency in lossless plasma. (This energy-part is partly dissipated, partly reflected to the lower atmosphere.) These two modes arise and propagate ahead completely independently of each other and this independence has no bearing upon the fact of longitudinal or oblique propagating (see part 2.5)

Further, the most essential new results – verified by the presented solution of Maxwell's equations – will be summarised, in contrast with the assumptions based on the former model-calculations (*Walker*, 1993; *Smith and Brice*, 1964; *Barrington et. al.*, 1966; *Gurnett et al.*, 1965; *Gurnett and Shawhan*, 1966; *Jones*, 1969; *Rodriguez and Gurnett*, 1971; *Terry and Budden*, 1978; *Leer et al.*, 1978; *Wang*, 1971; *Watanabe and Ondoh*, 1976; *Fijalkow et al.*, 1973; *Cserepes*, 1987; *Wong and Goldstein*, 1987; *Thorne and Horne*, 1994b; *Horne and Thorne*, 1994; *Villalón and Burke*, 1993; *Anderson and Fuselier*, 1994; *Singh et al.*, 1998):

- Proton-whistlers can occur in a plasma containing only protons and electrons, so there is no need and no reason to assume at least two different kinds of positive ions for generating proton whistlers – contrary to the earlier assumptions (eg. *Wang*, 1971; *Walker*, 1993).

- It is possible for the proton-whistlers to arise and propagate strictly longitudinally along the magnetic field and reach the satellite, therefore the longitudinal propagation is not anomalous from any aspect – contrary to the former hypotheses (*Wang*, 1971).

- The most fundamental difference is the analytical verification of the fact that no polarization-reversal happens at any so called cross-over frequency and the different modes are originally independent of each other – contrary to the former theories (*Walker*, 1993).

- Further, it became clear that proton-whistlers can propagate in a homogeneous and lossless medium as well as in an inhomogeneous one, thus the spatial inhomogeneity and the presence of loss are not preliminary conditions in the explanation of ion cyclotron wave-propagation – contrary to some former model-calculations (e.g. *Wang*, 1971; *Walker*, 1993).

- As was mentioned in part 2.2, the influence of loss is similar to the influence of ions, as the same part of the signal (EM-mode) starts to propagate in these cases. Obviously, the exact signal-forms are different (e.g. the cut-off frequency in lossy case is the electron gyro-frequency for both modes, but in case of ion-effect the EM-mode cut-off is at the proton gyro-frequency in lossless case, while the HM-mode cut-off remains at the electron gyro-frequency).

2.4 Whistler-mode solution in a plasma containing ions with arbitrary distribution (phenomenon of ion-cyclotron whistlers)

The presence of protons in a magnetised plasma will essentially modify the final form of the propagating whistler-mode signal, as was detailed in part 2.3. A more accurate description of the reality is available by considering a multi-component plasma with arbitrarily chosen ion-distribution. In this case, the magnetised neutral plasma (medium "2") contains electrons and n different kinds of positive particles, where

$$N = N^- = N^+ = \sum_{i=1}^{n} N_i^+ , \qquad (2.133)$$

and

$$\omega_{bi}^+ = \frac{q_i^+}{m_i^+} B_{F0} ; \; (\omega_{pi}^+)^2 = \frac{(q_i^+)^2 N_i^+}{\varepsilon_0 m_i^+} . \qquad (2.134)$$

2.4.1 Solution for a homogeneous plasma

Maxwell's equations are in the well-known form – see (1.24) –, however the signal-medium interaction is different as follows

$$
\begin{aligned}
&m^- \frac{\partial \bar{v}^-}{\partial t} = q^-(\bar{E} + \bar{v}^- \times \bar{B}_{F0}), && m_i^+ \frac{\partial \bar{v}_i^+}{\partial t} = q_i^+(\bar{E} + \bar{v}_i^+ \times \bar{B}_{F0}), \\
&\bar{J}^- = q^- N \bar{v}^-, && \bar{J}_i^+ = q_i^+ N_i \bar{v}_i^+, \\
&\nabla \bar{J}^- + \frac{\partial \rho^-}{\partial t} = 0; && \nabla \bar{J}_i^+ + \frac{\partial \rho_i^+}{\partial t} = 0; \\
&\sum_{i=1}^{n} \bar{J}_i^+ = \bar{J}^+ = \sum_{i=1}^{n} q_i^+ N_i^+ \bar{v}_i^+, \\
&\bar{J} = \bar{J}^+ + \bar{J}^-.
\end{aligned} \tag{2.135}
$$

Due to the assumed plane-wave form, and $\partial/\partial y = \partial/\partial z \equiv 0$ and $E_x = 0$ is obtained. The equation-system to be solved in time-space domain

$$
\begin{aligned}
\frac{\partial^2 E_y}{\partial x^2} = \frac{1}{c^2}\Bigg\{ &\sum_{i=1}^{n}\Bigg[\frac{(\omega_{pi}^+)^2}{\omega_{bi}^+}\int^{t} \frac{\partial E_y(\tau)}{\partial \tau}\omega_{bi}^+ \cos\omega_{bi}^+(t-\tau)\,d\tau + \\
&+(\omega_{pi}^+)^2 \int^{t} E_z(\tau)\omega_{bi}^+ \cos\omega_{bi}^+(t-\tau)\,d\tau\Bigg] + \\
&+\frac{(\omega_p^-)^2}{\omega_b^-}\int^{t} \frac{\partial E_y(\tau)}{\partial \tau}\omega_b^- \cos\omega_b^-(t-\tau)\,d\tau + \\
&+(\omega_p^-)^2 \int^{t} E_z(\tau)\omega_b^- \cos\omega_b^-(t-\tau)\,d\tau + \frac{\partial^2 E_y}{\partial t^2}\Bigg\}, \\
\frac{\partial^2 E_z}{\partial x^2} = \frac{1}{c^2}\Bigg\{ &\sum_{i=1}^{n}\Bigg[\frac{(\omega_{pi}^+)^2}{\omega_{bi}^+}\int^{t} \frac{\partial E_z(\tau)}{\partial \tau}\omega_{bi}^+ \cos\omega_{bi}^+(t-\tau)\,d\tau - \\
&-(\omega_{pi}^+)^2 \int^{t} E_y(\tau)\omega_{bi}^+ \cos\omega_{bi}^+(t-\tau)\,d\tau\Bigg] + \\
&+\frac{(\omega_p^-)^2}{\omega_b^-}\int^{t} \frac{\partial E_z(\tau)}{\partial \tau}\omega_b^- \cos\omega_b^-(t-\tau)\,d\tau - \\
&-(\omega_p^-)^2 \int^{t} E_y(\tau)\omega_b^- \cos\omega_b^-(t-\tau)\,d\tau + \frac{\partial^2 E_z}{\partial t^2}\Bigg\}.
\end{aligned} \tag{2.136}
$$

Introducing the common notations for transformed form of the initial conditions as earlier, let $e_{z0t}(s) = A(s)$ and $e'_{z0t}(s) = B(s)$ be and the transformed equations are

$$c^2p^2E_y(p,s)=\sum_{i=1}^{n}\left[\left(\omega_{pi}^{+}\right)^2\frac{s^2}{s^2+\left(\omega_{bi}^{+}\right)^2}E_y(p,s)+\left(\omega_{pi}^{+}\right)^2\omega_{bi}^{+}\frac{s}{s^2+\left(\omega_{bi}^{+}\right)^2}E_z(p,s)\right]+$$
$$+\left(\omega_{p}^{-}\right)^2\frac{s^2}{s^2+\left(\omega_{b}^{-}\right)^2}E_y(p,s)+\left(\omega_{p}^{-}\right)^2\omega_{b}^{-}\frac{s}{s^2+\left(\omega_{b}^{-}\right)^2}E_z(p,s)+$$
$$+s^2E_y(p,s),$$
$$c^2p^2E_z(p,s)=\sum_{i=1}^{n}\left[\left(\omega_{pi}^{+}\right)^2\frac{s^2}{s^2+\left(\omega_{bi}^{+}\right)^2}E_z(p,s)-\left(\omega_{pi}^{+}\right)^2\omega_{bi}^{+}\frac{s}{s^2+\left(\omega_{bi}^{+}\right)^2}E_y(p,s)\right]+$$
$$+\left(\omega_{p}^{-}\right)^2\frac{s^2}{s^2+\left(\omega_{b}^{-}\right)^2}E_z(p,s)-\left(\omega_{p}^{-}\right)^2\omega_{b}^{-}\frac{s}{s^2+\left(\omega_{b}^{-}\right)^2}E_y(p,s)+$$
$$+s^2E_z(p,s)+c^2[p\cdot A(s)+B(s)].$$

(2.137)

With the notations

$$u_i^{2+}=s^2+\left(\omega_{bi}^{+}\right)^2,$$
$$u^{2-}=s^2+\left(\omega_{b}^{-}\right)^2,$$

(2.138)

the following relations can be obtained for the transformed field-components

$$E_y(p,s)=c^2\frac{a_\infty(s)\cdot a_0(s)\cdot a_2(s)}{b_3(s)}\cdot\frac{p+\dfrac{a_1(s)}{a_2(s)}}{p^4-p^2\dfrac{b_2(s)}{b_3(s)}+\dfrac{b_1(s)}{b_3(s)}},$$
$$E_z(p,s)=c^2\frac{a_\infty(s)\cdot a_2(s)\cdot a_4(s)}{b_3(s)}\cdot\frac{\left[p+\dfrac{a_1(s)}{a_2(s)}\right]\left[p^2-\dfrac{a_3(s)}{a_4(s)}\right]}{p^4-p^2\dfrac{b_2(s)}{b_3(s)}+\dfrac{b_1(s)}{b_3(s)}},$$

(2.139)

where the polynomials are

$$
\begin{aligned}
&a_1(s)=B(s),\\
&a_2(s)=A(s),\\
&a_\infty(s)=u^{2-}\prod_{i=1}^{n}u_i^{2+},\\
&a_0(s)=s\left[u^{2-}\left(\sum_{i=1}^{n}(\omega_{pi}^{+})^2\omega_{bi}^{+}\prod_{j\neq i}u_j^{2+}\right)+(\omega_p^{-})^2\omega_b^{-}\prod_{i=1}^{n}u_i^{2+}\right],\\
&a_3(s)=s^2\left[u^{2-}\left(\sum_{i=1}^{n}(\omega_{pi}^{+})^2\prod_{j\neq i}u_j^{2+}\right)+(\omega_p^{-})^2\prod_{i=1}^{n}u_i^{2+}+u^{2-}\prod_{i=1}^{n}u_i^{2+}\right],\\
&a_4(s)=c^2u^{2-}\prod_{i=1}^{n}u_i^{2+},\\
&b_1(s)=s^4\left[u^{2-}\left(\sum_{i=1}^{n}(\omega_{pi}^{+})^2\prod_{j\neq i}u_j^{2+}\right)+(\omega_p^{-})^2\prod_{i=1}^{n}u_i^{2+}+u^{2-}\prod_{i=1}^{n}u_i^{2+}\right]^2+\\
&\qquad+s^2\left[u^{2-}\left(\sum_{i=1}^{n}(\omega_{pi}^{+})^2\omega_{bi}^{+}\prod_{j\neq i}u_j^{2+}\right)+(\omega_p^{-})^2\omega_b^{-}\prod_{i=1}^{n}u_i^{2+}\right]^2,\\
&b_2(s)=2c^2\left(\prod_{i=1}^{n}u_i^{2+}\right)u^{2-}s^2\left[u^{2-}\left(\sum_{i=1}^{n}(\omega_{pi}^{+})^2\prod_{j\neq i}u_j^{2+}\right)+(\omega_p^{-})^2\prod_{i=1}^{n}u_i^{2+}+u^{2-}\prod_{i=1}^{n}u_i^{2+}\right],\\
&b_3(s)=c^4\left(\prod_{i=1}^{n}u_i^{2+}\right)^2(u^{2-})^2,
\end{aligned}
\tag{2.140}
$$

The four poles of the denominator are

$$
\begin{aligned}
&p_{1,2,3,4}=\pm\frac{1}{c}\sqrt{\frac{NUM}{DEN}};\\
&NUM=s^2\left\{u^{2-}\sum_{i=1}^{n}\left[(\omega_{pi}^{+})^2\prod_{j\neq i}u_j^{2+}\right]+(\omega_p^{-})^2\prod_{i=1}^{n}u_i^{2+}+u^{2-}\prod_{i=1}^{n}u_i^{2+}\right\}\pm\\
&\qquad\pm js\left\{u^{2-}\sum_{i=1}^{n}\left[(\omega_{pi}^{+})^2\omega_{bi}^{+}\prod_{j\neq i}u_j^{2+}\right]+(\omega_p^{-})^2\omega_b^{-}\prod_{i=1}^{n}u_i^{2+}\right\},\\
&DEN=u^{2-}\prod_{i=1}^{n}u_i^{2+}.
\end{aligned}
\tag{2.141}
$$

Executing the $s=j\omega$, $u_i^{2+}=s^2+(\omega_{bi}^{+})^2$, $u^{2-}=s^2+(\omega_b^{-})^2$ substitutions and taking into consideration that $\omega_b^{-}<0\,(|\omega_b^{-}|\triangleq\omega_{be})$, the form of the poles is

$p_{1,2,3,4} =$

$$= \pm \frac{1}{c}\sqrt{\frac{-\omega^2\left[\left(\omega_p^-\right)^2+\omega_{be}^2-\omega^2\right]\pm\omega\left(\omega_p^-\right)^2\omega_{be}}{\omega_{be}^2-\omega^2}+\left(-\omega^2\right)\sum_{i=1}^{n}\frac{\left(\omega_{pi}^+\right)^2}{\left(\omega_{bi}^+\right)^2-\omega^2}\mp\omega\sum_{i=1}^{n}\frac{\left(\omega_{pi}^+\right)^2\omega_{bi}^+}{\left(\omega_{bi}^+\right)^2-\omega^2}} \tag{2.142}$$

The two forward propagating modes belong to p_1 and p_3 – in concordance with the former model calculations. (The backward propagating modes belonging to p_2 and p_4 are negligible as the magnetosphere i.e. the half-space "2" is supposed to be an infinite half-space.)

The p_1 pole is clearly imaginary in the $0<\omega<\omega_{be}$ range

$$p_1 = -\mathrm{j}\frac{1}{c}\sqrt{\frac{\omega^2\left[\left(\omega_p^-\right)^2+\omega_{be}^2-\omega^2\right]+\omega\left(\omega_p^-\right)^2\omega_{be}}{\omega_{be}^2-\omega^2}+\omega^2\sum_{i=1}^{n}\frac{\left(\omega_{pi}^+\right)^2}{\left(\omega_{bi}^+\right)^2-\omega^2}-\omega\sum_{i=1}^{n}\frac{\left(\omega_{pi}^+\right)^2\omega_{bi}^+}{\left(\omega_{bi}^+\right)^2-\omega^2}} = -\mathrm{j}k_1 \tag{2.143}$$

therefore the mode belonging to p_1 is the electron-whistler. The p_3 pole can change its nature according to ω. While in $\omega_{bi\,\max}^+<\omega<\omega_{be}$ there is no possibility of propagation of the mode belonging to p_3 in this lossless model, in $0<\omega<\omega_{bi\,\max}^+$ this mode can propagate (depending on the actual value of $\Sigma_{i=1}^{n}$ "ion-terms"). This mode will deliver the so-called ion-cyclotron whistlers

$$p_3 = -\mathrm{j}\frac{1}{c}\sqrt{\omega^2\sum_{i=1}^{n}\frac{\left(\omega_{pi}^+\right)^2}{\left(\omega_{bi}^+\right)^2-\omega^2}+\omega\sum_{i=1}^{n}\frac{\left(\omega_{pi}^+\right)^2\omega_{bi}^+}{\left(\omega_{bi}^+\right)^2-\omega^2}-\frac{\omega\left(\omega_p^-\right)^2\omega_{be}+\omega^4-\omega^2\left[\left(\omega_p^-\right)^2+\omega_{be}^2\right]}{\omega_{be}^2-\omega^2}} = -\mathrm{j}k_3,$$

if $0<\omega<\omega_{bi\,\max}^+$, and **(2.144)**

$$p_3 = -\frac{1}{c}\sqrt{\frac{-\omega^2\left[\left(\omega_p^-\right)^2+\omega_{be}^2-\omega^2\right]+\omega\left(\omega_p^-\right)^2\omega_{be}}{\omega_{be}^2-\omega^2}-\omega^2\sum_{i=1}^{n}\frac{\left(\omega_{pi}^+\right)^2}{\left(\omega_{bi}^+\right)^2-\omega^2}-\omega\sum_{i=1}^{n}\frac{\left(\omega_{pi}^+\right)^2\omega_{bi}^+}{\left(\omega_{bi}^+\right)^2-\omega^2}} = -\alpha_3,$$

if $\omega_{bi\,\max}^+<\omega<\omega_{be}$. This fact demonstrates that the different ion-cyclotron whistlers generated by different kinds of ions are not independent of each other, but there is only one mode existing, including all the influences of ions.

From this step, the solution is identical with the one presented earlier for protons. The electric field components for electron- and ion-cyclotron whistlers are

$$E_{2zew} = -\frac{Z_0}{4\pi} \int_{-\omega_{max}}^{\omega_{max}} I_{x0}(\omega) \frac{k_0(\omega)}{k_0(\omega) + k_1(\omega)} e^{j[\omega t - k_1(\omega)(x - x_0)]} d\omega,$$
$$E_{2ziw} = -\frac{Z_0}{4\pi} \int_{-\omega_{max}}^{\omega_{max}} I_{x0}(\omega) \frac{k_0(\omega)}{k_0(\omega) + k_3(\omega)} e^{j[\omega t - k_3(\omega)(x - x_0)]} d\omega, \qquad (2.145)$$

if $0 < \omega < \omega^{+}_{bi\,max}$, and

$$E_{2zew} = -\frac{Z_0}{4\pi} \int_{-\omega_{max}}^{\omega_{max}} I_{x0}(\omega) \frac{k_0(\omega)}{k_0(\omega) + k_1(\omega)} e^{j[\omega t - k_1(\omega)(x - x_0)]} d\omega,$$
$$E_{2zN} = \frac{Z_0}{4\pi} \int_{-\omega_{max}}^{\omega_{max}} I_{x0}(\omega) \frac{k_0(\omega)}{j\alpha_3(\omega) - k_0(\omega)} e^{j\omega t - \alpha_3(\omega)(x - x_0)} d\omega, \qquad (2.146)$$

if $\omega^{+}_{bi\,max} < \omega < \omega_{be}$. The form of excitation is identical with the one in equation (1.67).

2.4.2 Solution for a weakly inhomogeneous plasma

The spatial inhomogeneity in the plasma is taken into account by the plasma parameters depending on space

$$\omega^{+}_{bi}(x),\ \omega^{+}_{pi}(x),\ \omega^{-}_{b}(x),\ \omega^{-}_{p}(x),\ N(x),\ \overline{B}_{F0}(x). \qquad (2.147)$$

The effectively applied ion-distribution for describing the inhomogeneity will be shown in the next part.

As, apart from small details, the course of computation in the present case does not differ from the previous ones, the final form of whistler mode solution in multi-component, weakly inhomogeneous plasma is as follows

$$E_{2zew}(x,t)=-\frac{Z_0}{4\pi}\int_{-\omega_{max}}^{\omega_{max}} I_{x0}(\omega)\sqrt{\frac{k_1(x_0,\omega)}{k_1(x,\omega)}}\frac{k_0(\omega)}{k_0(\omega)+k_1(x_0,\omega)}\cdot e^{j\left[\omega t-\int_{x_0}^{x}k_1(\xi,\omega)d\xi\right]}d\omega,$$

$$H_{2yew}(x,t)=\frac{1}{4\pi}\int_{-\omega_{max}}^{\omega_{max}} I_{x0}(\omega)\sqrt{\frac{k_1(x,\omega)}{k_1(x_0,\omega)}}\frac{k_1(x_0,\omega)}{k_0(\omega)+k_1(x_0,\omega)}\cdot e^{j\left[\omega t-\int_{x_0}^{x}k_1(\xi,\omega)d\xi\right]}d\omega;$$

$$E_{2ziw}(x,t)=-\frac{Z_0}{4\pi}\int_{-\omega_{max}}^{\omega_{max}} I_{x0}(\omega)\sqrt{\frac{k_3(x_0,\omega)}{k_3(x,\omega)}}\frac{k_0(\omega)}{k_0(\omega)+k_3(x_0,\omega)}\cdot e^{j\left[\omega t-\int_{x_0}^{x}k_3(\xi,\omega)d\xi\right]}d\omega,$$

$$H_{2yiw}(x,t)=\frac{1}{4\pi}\int_{-\omega_{max}}^{\omega_{max}} I_{x0}(\omega)\sqrt{\frac{k_3(x,\omega)}{k_3(x_0,\omega)}}\frac{k_3(x_0,\omega)}{k_0(\omega)+k_3(x_0,\omega)}\cdot e^{j\left[\omega t-\int_{x_0}^{x}k_3(\xi,\omega)d\xi\right]}d\omega,$$

(2.148)

where $k_0=\omega/c$, $I_{x0}(\omega)$ is given by equation (1.67) and for (k_1,k_3) see (2.143) and (2.144)

2.4.3 Numerical results

The IRI (International Reference Ionosphere, *Bilitza*, 1990) model of the electron- and ion-density distributions considered in the multi-component model-calculations can be seen in Fig. 2.34 and Fig. 2.35.

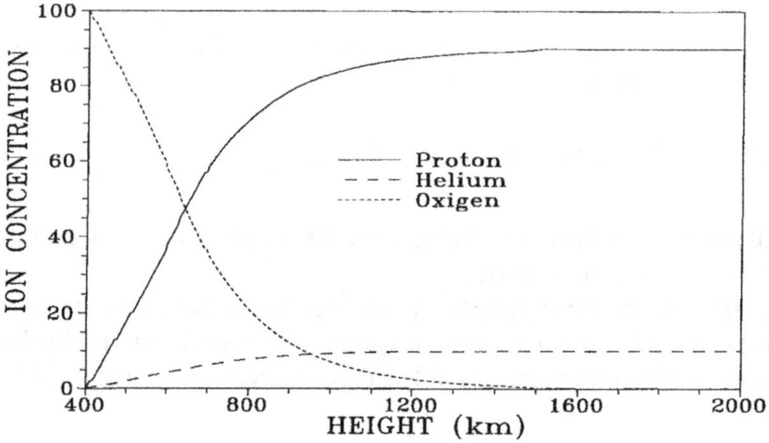

Figure 2.35. Ion-density distributions based on the IRI model

The frequency- and height-dependence of the propagating vector belonging to the ion-cyclotron mode (the theoretical possibilities of the propagation) are shown in Fig. 2.36 The dark areas represent the "cut-off"

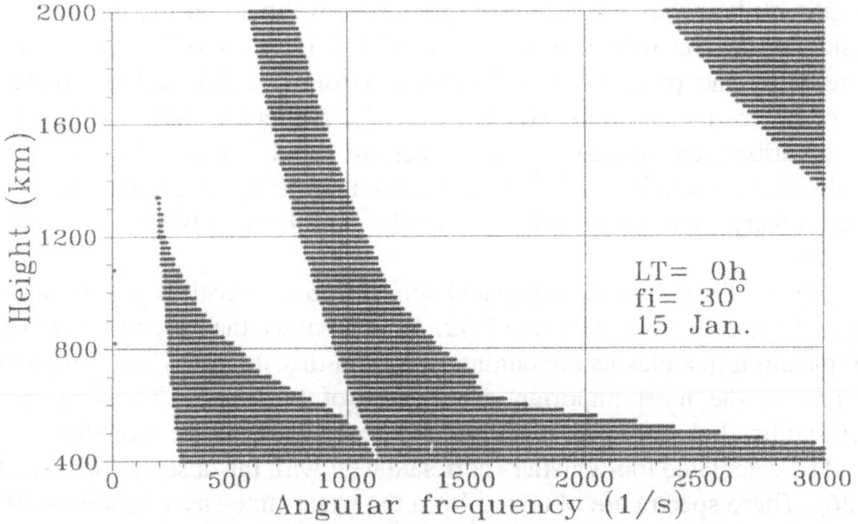

Figure 2.36. Propagation-domains for the ion-cyclotron mode

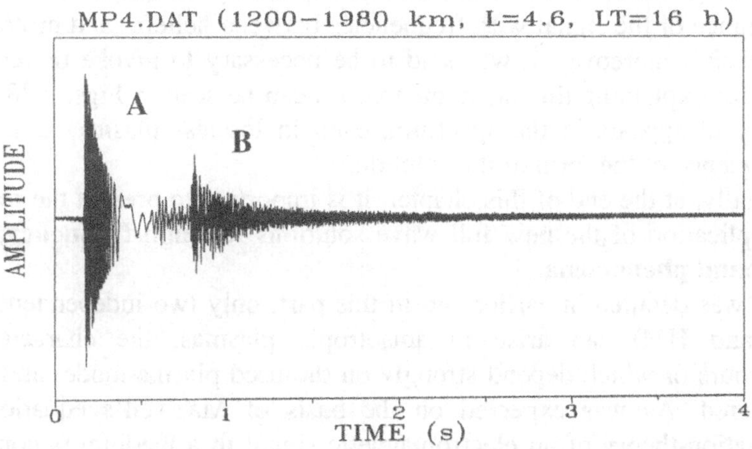

Figure 2.37. Time-function of a complete signal propagating in multi-component plasma

ranges, where the ion-cyclotron mode cannot propagate. The propagation is theoretically possible in the white domains.

Fig. 2.37 shows the time-function of a complete signal (electron- and ion-cyclotron-whistler together) in inhomogeneous multi-component plasma.

The first part (*A*) of the signal is the electron-whistler, and the ion-cyclotron mode appear from ~ 0.3s. This second part (*B*) is disintegrated into several amplitude-branches belonging to one mode in the spectrum, the first is the influence of the oxygen-ions on the signal (from ~ 0.3s), the second is caused by the presence of helium-ions (from ~ 0.5s), and the third one belongs to the protons (from ~ 0.7s). Although these parts of the observations are called oxygen-, helium- and proton-whistlers in the traditional terminology (as they were assumed to be modes independent of each other), they form only one mode (left hand rotating ion-cyclotron mode).

The FFT-patterns of calculated ion-cyclotron-whistlers can be seen in Figs. 2.38, 2.39, 2.40, 2.41 and 2.42. As is obvious, the starting height of the propagation in a plasma containing given ion-distributions ("entering point") is one of the most important parameters of the signal. The first, nearly vertical line belongs to the electron-whistler in the dynamic spectrum.

The calculated ion-whistlers correlate well with the observations (see Fig. 2.26). These spectra are obtained from the exact time-space functions of the complete signals. This result basically, qualitatively differs from any other model-spectrum calculated from monochromatic approximations, and opens the way to analyse the fine structure of these phenomena – even the type of the excitation and the actual ion-concentrations. For example, in *Walker*'s approach (*Walker*, 1993), it was hard to explain by the traditional theory the appearance of the signal with frequencies between helium- and proton-gyro-frequencies, moreover, it was said to be necessary to invoke the collision-effect for explaining this phenomenon. As can be seen in Figs. 2.38 ~ 2.42, this signal appears in the spectrum, even in lossless plasma, as a natural consequence of the form of the solution.

Finally, at the end of this chapter, it is important to present the results of the application of the new full wave solutions for **high frequency, extra-wide band phenomena**.

As was detailed in earlier and in this part, only two independent modes (EM and HM) can arise in anisotropic plasmas, the characters and behaviours of which depend strongly on the used plasma-model and effects considered. As was expected on the basis of Maxwell's equations, the propagation-theory of an electromagnetic signal in a medium is consistent, and the validity of an accurate solution is not restricted by the frequency-band investigated. Therefore, the presented two modes must be the only possible existing modes in the complete frequency-range, appearing and propagating always independently of each other. A consequence of this fact is the theoretical generality of the obtained formulas.

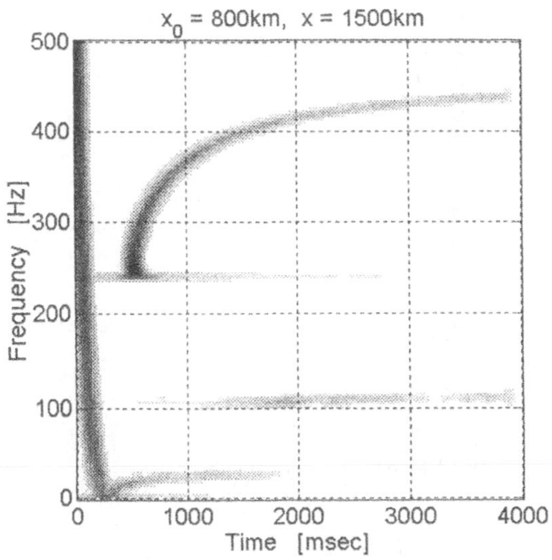

Figure 2.38. FFT-pattern of an ion-whistler entering the plasma at 800km

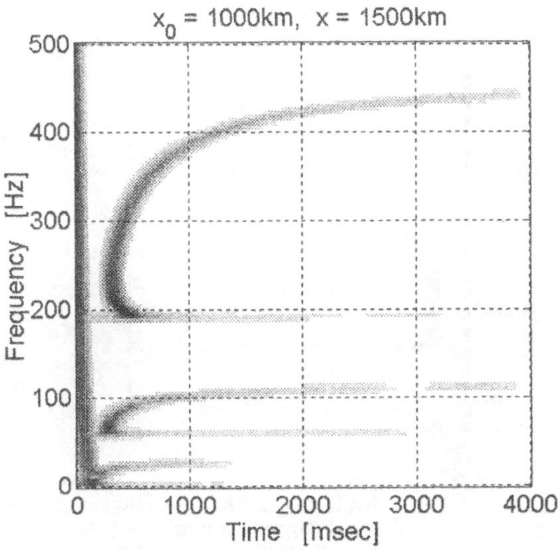

Figure 2.39. FFT-pattern of an ion-whistler entering the plasma at 1000km

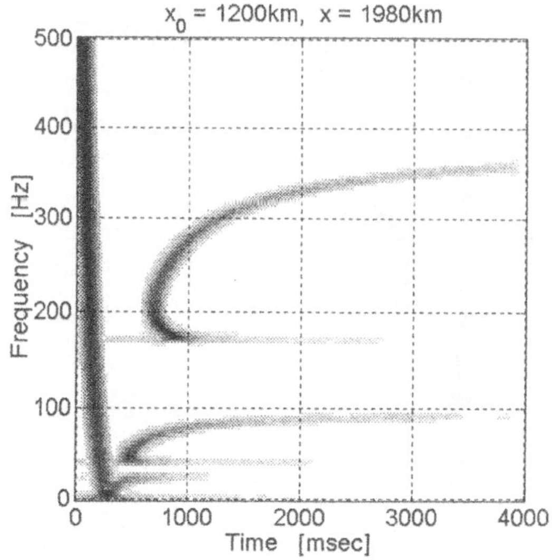

Figure 2.40. FFT-pattern of an ion-whistler entering the plasma at 1200km

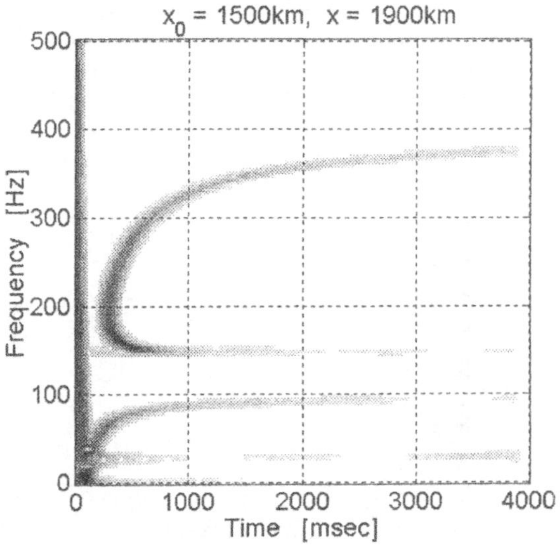

Figure 2.41. FFT-pattern of an ion-whistler entering the plasma at 1500km

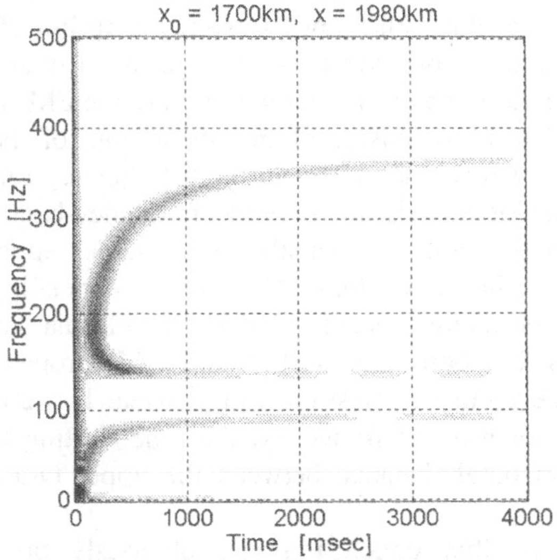

Figure 2.42. FFT-pattern of an ion-whistler entering the plasma at 1700km

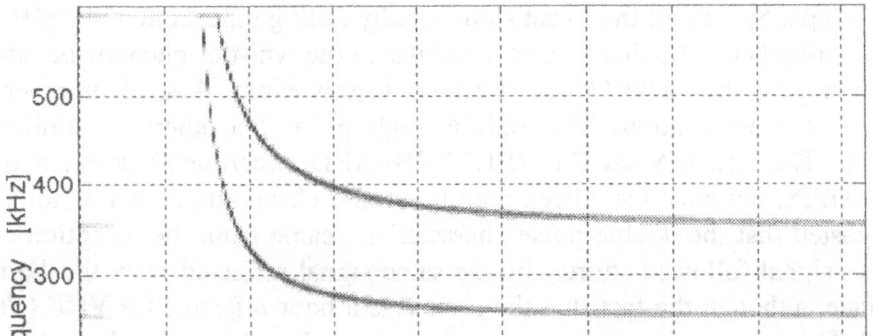

Figure 2.43. FFT-pattern of the full signal containing the ELF-VLF and the high frequency terms

In the case of $\omega \gg \omega_{pe}, \omega_{be}$ (high-frequency bands) the presented solution delivers e.g. the well-known **Faraday rotation** effect (the plasma decomposes the plane-wave into left- and right-hand rotating EM and HM modes with different propagation-factor, therefore the EM-mode falls behind the HM-mode). By the extension of the upper limit of the integral in the (propagation-factor) formulas up to the high-frequency ranges (MHz), the dynamic spectrum of the signal for wide frequency-band is presented in Fig.2.43. As can be seen, the whistler (HM mode) appears in the low-frequency domain. The ion-cyclotron (EM) mode, depending on the plasma-parameters, can or cannot penetrate into the plasma and propagate in ELF~VLF bands, but both modes (EM and HM) propagate in the high-frequency ranges even in a lossless electron-plasma. This effect results in the Faraday rotation for a given frequency, the time-shifting between the two modes is the horizontal distance between the upper two branches of the spectrum.

The validity of this conclusion can obviously be proved by the investigation of the so-called **TiPP-events** (*Rodger*, 1999). TiPPs (**T**rans**i**onospheric **P**ulse **P**airs) were first reported in 1995 (*Holden et al.*, 1995; *Massey and Holden*, 1995). They are satellite-observed electromagnetic bursts with duration of approximately 10 μs, are typically 20-40 dB stronger than the background, and occur in pairs separated by ~ 10-110 μs. Spectra of the event show rapidly falling tone from >140 MHz to a few megahertz. As this feature is similar to the whistler-phenomena, some experts gave the name "*Megawhistlers*" to this event. A small number (~ 1%) of the observations have only a single pulse, but otherwise similar to TiPPs. The ALEXIS satellite (BLACKBEARD experiment) detected over 850 TiPPs, but none have been seen in ground observations. It was initially suggested that the double-pulse characteristic came from the reception of a direct signal followed shortly by the same signal reflected from the Earth's surface, although the fact that the ground is a poor reflector for VHF (**V**ery **H**igh **F**requency) signals is commonly known. It is becoming increasingly clear that some TiPPs are associated with thunderstorms, but any explanation of the physical origin of the phenomena or any description of the signal have been missed up to now. An observed TiPP-event can be seen in Fig. 2.44.

The open question of the origin and the shape of these phenomena can be answered easily and naturally by the aid of the new full-wave solutions of the Maxwell's equations. As was mentioned above, both the EM and the HM modes propagate in the VHF-range. This causes the spectral splitting in both detected traces, as they contain actually two (left- and right hand rotating) modes. This can be seen in Fig. 2.45. Further, the reason why these phenomena appear nearly always in pairs is the shape of the excitation.

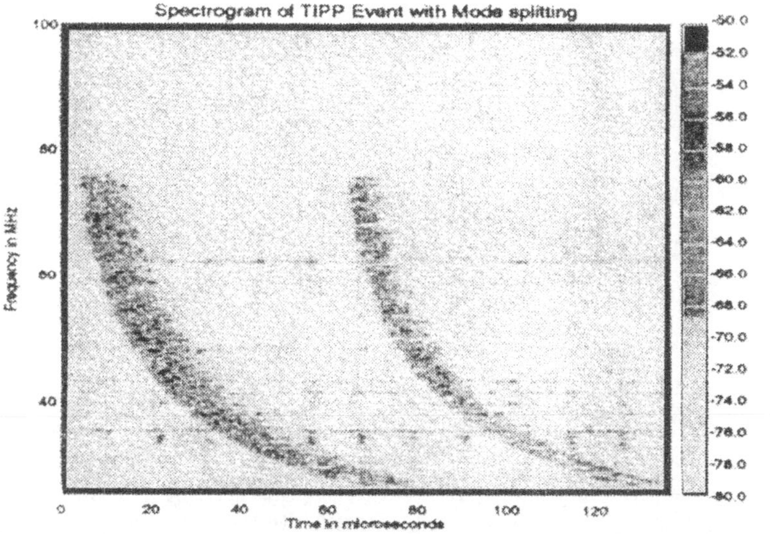

*Figure 2.44.*Spectrogram of a TiPP event taken from the BLACKBEARD experiment on board the ALEXIS satellite (After *Holden et al.*, 1995, Fig. 3)

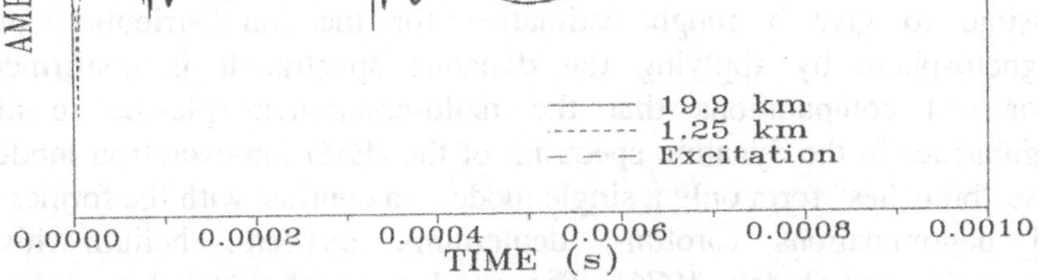

Figure 2.45a. Time function of calculated model-whistler close to the square impulse excitation

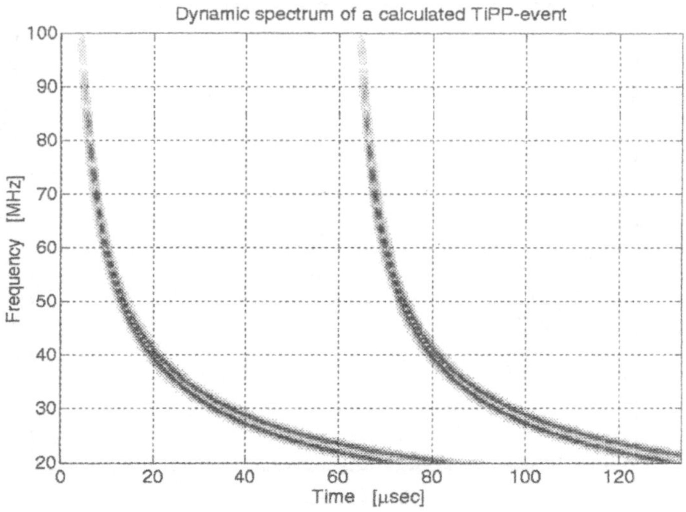

Figure 2.45b. Calculated TiPP event obtained from the full-wave model, for a 50 μs long square impulse excitation

The character of a lightning stroke can most frequently be described by a square impulse, containing an upward and a downward "directed" edge. The substitution of this function into the excitation (I_{x0}) in the full-wave model results automatically in the spectrum shown in Fig. 2.45. The calculated and the measured signals correlate very well.

2.4.4 Conclusions

The extension of the solution is successful for plasma-models containing arbitrary ion-distributions.

On the basis of the solution presented in closed form, it seems to be possible to give a rough estimation for the ion-distribution of the magnetosphere by applying the dynamic spectra. It is confirmed by theoretical computations that the multi-component plasma results in singularities in the dynamic spectrum of the (EM) ion-cyclotron mode, but these "branches" form only a single mode – in contrast with the former ideas and denominations (proton-, deuterium-, oxygen-, helium-whistlers) (*Watanabe and Ondoh*, 1976). The conclusions of part 2.3 regarding the former assumptions are valid in the case of multi-component plasma as well (see e.g. polarization-reversal).

The phenomenon in the VLF-band is analogous with the one in HF-VHF-range, which is known as Faraday rotation for monochromatic plane-waves.

As is known, the explanation of Faraday rotation does not contain polarization-reversal or mode coupling either. Moreover, the phenomenon of TiPPs can be unambiguously explained by the results coming from the full-wave model. This fact verifies the authenticity of the results, furthermore, creates the continuity of the theory of electromagnetic propagational phenomena, answering numerous open questions and delivering more exact descriptions instead of the former, approximate hypotheses.

2.5 Whistler-mode solution in case of non-longitudinal propagation

2.5.1 Introduction

The problem to be examined in this chapter is an essential and up to the present open question of the whistler-investigation – involving the theory of ducts. Numerous theories contain the assumption of existence of ducting channels along the geomagnetic field-lines in which the whistler-mode signals propagate from one hemisphere to the other. Many observed effects are explained by the existence of these channels in the magnetosphere, and their spatial and temporal changes. *Storey's* results (1953) opened the history of field-aligned propagation and the possibility to analyse the ionosphere-magnetosphere structure by the aid of whistlers and started world-wide extended research-work (*Brice and Smith*, 1971; *Sazhin et al.*, 1992a). The theory of field-aligned propagation was supported by measurements at the conjugate field-line points (*Morgan and Allcock*, 1956), and by investigation of terrestrial radio transmitters of 155 kHz (*Helliwell and Gehrels*, 1958). The first observations above the ionosphere were detected by a proton-precession magnetometer of the satellite Vanguard III. (*Cain et. al.*, 1962). Ray-tracing computations (*Maeda and Kimura*, 1956; *Yabroff*, 1961) assuming a monotone and continuous variation of the electron-density profile have shown a significant deviation of the wave normal vectors of whistlers from the geomagnetic field line, so it seemed to be impossible to receive them at the conjugate points. So arose the claim to suppose the existence of some kind of ducting channels ("ducts" of enhanced or depleted electron-density) along magnetic field lines. The detailed ray-optical approximation of the ducted propagation was carried out by *Smith* (1960) and *Smith et. al.* (1960). Later, *Scarabucci* and *Smith* (1971), *Walker* (1976), *Laird and Nunn* (1975) examined the possibility of whistler-mode propagation in duct-models having plan-parallel or circular cross sections (with numerous simplifications in the models). *Strangeways* (1977, 1982) and others executed ray-tracing model-calculations in the case of ducting channels containing different electron-density distributions. Nevertheless, these theories are based on some approximate restrictions – e.g. the monochromatic-type ray-tracing – and do not deliver sufficient answer for some measured phenomena – e.g. the *L*-discrepancy. Beyond the duct-theory worked out for terrestrial observations, a fundamentally different phenomenon was recognised by the satellite measurements, which is named *L*-discrepancy (*Tarcsai*, 1981). It became possible to identify the whistlers measured on the ground and at the satellite simultaneously. In the analysis of this type of data (e.g. 25.03.1974, IK-10 satellite and Panská Ves terrestrial station) the signals detected were separated into two groups: "ducted" type of signals (can be seen in both observations) and "non-ducted" type (can be seen only in satellite observations). It could be expected that the ducted whistlers propagated along the field-line crossed by the satellite at the moment of registration. Instead of this, there is a systematic difference between the *L*-values estimated by evaluating the whistlers and the real satellite position. This confirms the fact of non-longitudinal propagation in these cases, and opens a number of questions concerning duct theory. Two explanations exist for the *L*-discrepancy. In *Cerisier's* approach (*Tarcsai*, 1981) the whistlers leave the ducts a few hundred kilometres far above the satellite and radiate/scatter within a cone, in which their pattern and travel-time are essentially unchanged. In *Thomson's* (*Thomson*, 1977a,b) explanation the whistlers reach the satellite after reflecting upward from the ionosphere. The most conspicuous contradiction in these theories is caused by the fact that these signals are detectable at terrestrial stations, however, they would not be ducted along their path, as these whistlers observed by satellites must propagate non-longitudinally to the point of registration. Furthermore, the signals that propagate along different paths in dispersive, anisotropic medium show significant differences in their spectra, so the assumption of any kind of unvarying pattern must be false. So arose the claim to investigate the scattered (non-ducted) whistlers.

In the present model, the possibility of propagation having different direction from the geomagnetic field line will be examined in case of *homogeneous* plasma. The most important questions are: what kind of modification may occur in the shape of the whistler and in its spectrum, if the geomagnetic field and the direction of propagation is not parallel to each other? Is it possible to determine unambiguously the actual way of propagation on the basis of terrestrial or satellite measured data?

2.5.2 The applied model

The primary limitation of the applicable model is the mathematical complexity. As can be seen in Fig. 2.46, assuming the superimposing magnetic dipole-field, the instantaneous value of the angle between the magnetic field and the propagation factor is depending on space. This fact results in inhomogeneity in the plasma-model.

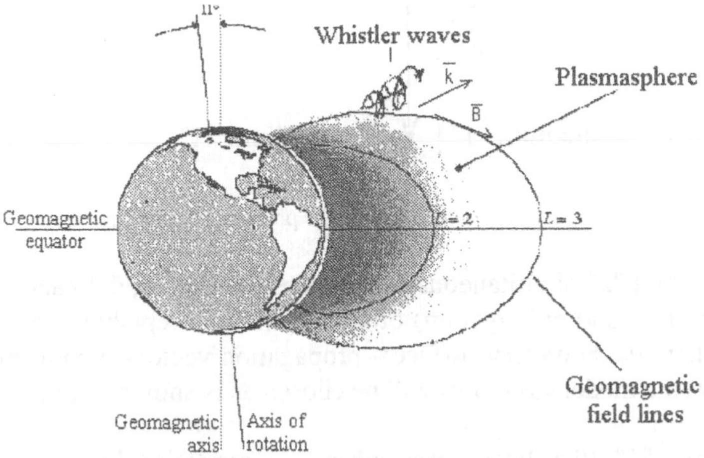

Figure 2.46. Oblique propagation in the magnetosphere

The form of the differential equations would be essentially modified and overcomplicated for non-monochromatic (impulse-like) signal investigated in previous models. This kind of inhomogeneity is not allowed to be approximated by the common quasi-homogeneous methods. Because of this, homogeneous plasma will be assumed in the presented model. Keeping the plane-wave assumption in the investigation, a fundamental theoretical contradiction appears concerning the geometry of the model. The problem can be seen in Fig. 2.47.

If the excitation is an infinite impulse-plane of arbitrarily directed propagation (as the shape of the signal is plane-wave having definite temporal and spatial limits) and the boundary surface is also an infinite plane, the excitation will intersect the boundary surface at the $t = 0$ moment, at a given point or segment. This fact results in not only a physical contradiction – as this phenomenon does not occur in the formation of whistlers in the Earth's atmosphere –, but opens wave-propagation theoretical questions about this model – as the source of signal is present in

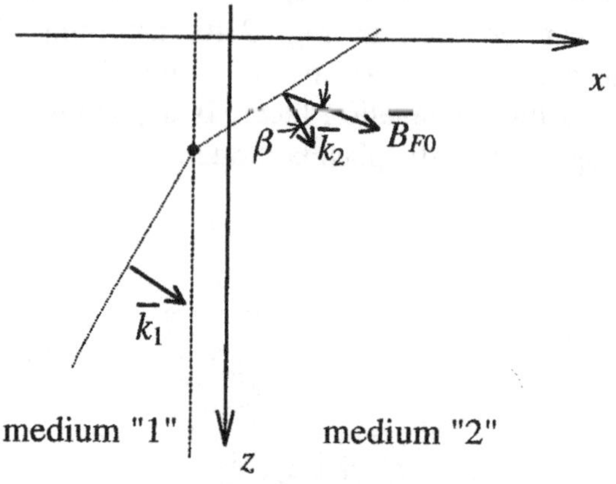

Figure 2.47. Illustration to the problem of coupling

media "1" and "2" simultaneously and therefore this model cannot describe the generation (and propagation) of signals in an acceptable form. To avoid this problem, the boundary surface – propagation vector – geomagnetic field geometry (the model structure) will be chosen as is shown in Fig. 2.48 a) and b).

Medium "1" (the lower atmosphere) is modelled by vacuum, while medium "2" (the plasma-sphere) is assumed to be cool, time-invariant, lossless, anisotropic, homogeneous electron-plasma. It is obvious that – in concordance with the literature – $\partial/\partial y \equiv 0$, given from the cylindrical symmetry of the model-structure (see Fig. 2.48a and b). The necessity of the coordinate-transformation form I. to II. is arisen from the mathematical complexity of the equations to be solved. If the direction of propagation is perpendicular to the boundary surface, and the superimposing magnetic field is of arbitrarily chosen direction, the problem outlined above is eliminated. On the other hand, two possibilities are given concerning the model of the plasma, depending on the choice of $\bar{e}_x$.

If $\bar{e}_x$ is parallel to $\bar{k}$, and $B_{F0} = B_{Fx} \cdot \bar{e}_x + B_{Fz} \cdot \bar{e}_z$, where

$$
\begin{aligned}
&|\bar{B}_{F0}| = B_{F0} \\
&B_{Fx} = B_{F0} \cdot \cos\beta; \quad B_{Fz} = B_{F0} \cdot \sin\beta.
\end{aligned}
\qquad \textbf{(2.149)}
$$

The form of the differential equations delivering the velocity-components will be the following

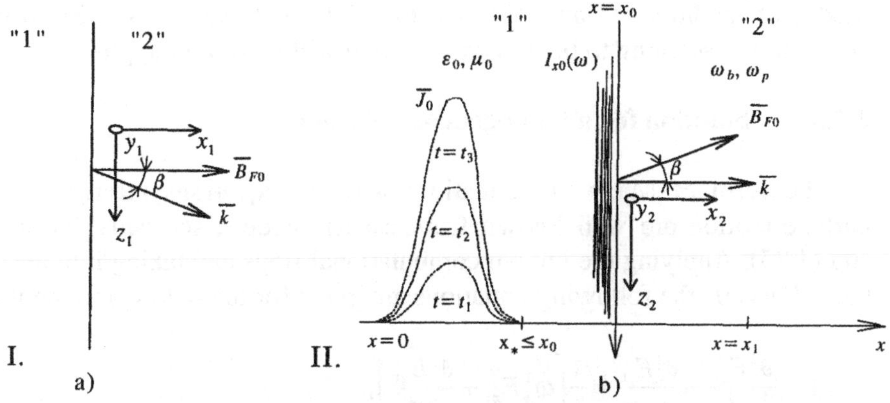

Figure 2.48. The coordinate-transformation

$$
\begin{aligned}
\text{I.}\quad & \frac{\partial^4 v_x}{\partial x^4} + 2\left(\frac{q}{m} B_{F0} \sin\beta\right)^2 v_x = \frac{\partial^3 E_x}{\partial x^3} + \left(\frac{q}{m} B_{F0} \sin\beta\right)\frac{\partial^2 E_y}{\partial x^2} + \\
& \qquad + \left(\frac{q}{m} B_{F0} \cos\beta\right)\frac{\partial E_x}{\partial x} + \left(\frac{q}{m} B_{F0}\right)^2 \sin\beta\cos\beta \frac{\partial E_z}{\partial x}, \\
\text{II.}\quad & \frac{\partial^2 v_y}{\partial x^2} + \left(\frac{q}{m} B_{F0}\right)(\sin^2\beta - \cos^2\beta) v_x = \frac{\partial E_y}{\partial x} - \left(\frac{q}{m} B_{F0} \sin\beta\right) E_x + \\
& \qquad + \left(\frac{q}{m} B_{F0} \cos\beta\right) E_z, \\
\text{III.}\quad & v_z = \frac{1}{\left(\frac{q}{m} B_{F0}\right)^2 \sin\beta\cos\beta} \frac{\partial^2 v_x}{\partial x^2} + \operatorname{tg}\beta \cdot v_x - \\
& \qquad - \frac{1}{\left(\frac{q}{m} B_{F0}\right)^2 \sin\beta\cos\beta} \frac{\partial E_x}{\partial x} - \frac{1}{\frac{q}{m} B_{F0} \cos\beta} E_y .
\end{aligned}
\tag{2.150}
$$

With this choice of the coordinate-axes, the equations relating to the velocity-components are much more complicated than in the case of longitudinal propagation. The practicability and successfulness of this way are questionable.

The other possible way of modelling is to choose $\bar{e}_x$ parallel to $\bar{B}_{F0}$. In this case $\bar{k} = k_x \cdot \bar{e}_x + k_z \cdot \bar{e}_z$ (see Fig. 2.48b). A great advantage of this choice is that the form of the starting differential-equation system is not modified, but the rotation of the coordinate-axes will be necessary before coupling the modes of medium "1" and "2" (see Fig. 2.48b). Nevertheless, this way of computation seeming to be less complicated will be further applied.

2.5.3 Solution for a homogeneous plasma

The form of Maxwell's equations and the equations referring to the particle-motion are well known from earlier models, see equations (1.24) and (1.25). Applying the known computational steps and taking into account that $\partial/\partial y \equiv 0$, the following equations are given for the field-components

$$\text{I.}\quad \frac{\partial^2 E_{x1}}{\partial z^2} - \frac{\partial^2 E_{z1}}{\partial x \partial z} = \frac{1}{c}\left[\omega_p^2 E_{x1} + \frac{\partial^2 E_{x1}}{\partial t^2}\right],$$

$$\text{II.}\quad \frac{\partial^2 E_{y1}}{\partial z^2} + \frac{\partial^2 E_{y1}}{\partial x^2} = \frac{1}{c^2}\left[\omega_p^2 \int^t \frac{\partial E_{y1}}{\partial \tau}\cos\omega_b (t-\tau)\mathrm{d}\tau + \right.$$
$$\left. + \omega_b \omega_p^2 \int^t E_{z1} \cos\omega_b (t-\tau)\mathrm{d}\tau + \frac{\partial^2 E_{y1}}{\partial t^2}\right], \qquad \textbf{(2.151)}$$

$$\text{III.}\quad \frac{\partial^2 E_{z1}}{\partial x^2} + \frac{\partial^2 E_{x1}}{\partial x \partial z} = \frac{1}{c^2}\left[\omega_p^2 \int^t \frac{\partial E_{z1}}{\partial \tau}\cos\omega_b (t-\tau)\mathrm{d}\tau - \right.$$
$$\left. - \omega_b \omega_p^2 \int^t E_{y1} \cos\omega_b (t-\tau)\mathrm{d}\tau + \frac{\partial^2 E_{z1}}{\partial t^2}\right].$$

As can be seen from (2.151), three independent variables will be required in the Laplace-transformation

$$x_1 \overset{\mathcal{L}}{\longleftrightarrow} p,\ z_1 \overset{\mathcal{L}}{\longleftrightarrow} l,\ t \overset{\mathcal{L}}{\longleftrightarrow} s, \qquad \textbf{(2.152)}$$

$$F(p,l,s) = \iiint_0^\infty f(x,z,t) \cdot \mathrm{e}^{-px} \cdot \mathrm{e}^{-lz} \cdot \mathrm{e}^{st}\, \mathrm{d}x\, \mathrm{d}z\, \mathrm{d}t .$$

The initial values given for the field-components (denoted here by e_i) are $(i = 1,2,3 \sim x,y,z)$

$$
\begin{aligned}
&e_i(0,0,t) \xleftrightarrow{\mathcal{L}} e_{i00t}(s), && \left.\frac{\partial e_i(x,z,t)}{\partial x}\right|_{x=0} \xleftrightarrow{\mathcal{L}} e'_{i0zt}(l,s),\\
&e_i(0,z,t) \xleftrightarrow{\mathcal{L}} e_{i0zt}(l,s), && \left.\frac{\partial e_i(x,z,t)}{\partial z}\right|_{z=0} \xleftrightarrow{\mathcal{L}} e'_{ix0t}(p,s),\\
&e_i(x,0,t) \xleftrightarrow{\mathcal{L}} e_{ix0t}(p,s), && \\
&e_i(x,z,0) \xleftrightarrow{\mathcal{L}} e_{ixz0}(p,l), && \left.\frac{\partial e_i(x,z,t)}{\partial t}\right|_{t=0} \xleftrightarrow{\mathcal{L}} e'_{ixz0}(p,l).
\end{aligned} \tag{2.153}
$$

The transformed form of (2.151) is as follows

$$
\begin{aligned}
&l^2 E_{x1}(p,l,s) - l\, e_{1x0t}(p,s) - e'_{1x0t}(p,s) - p l E_{z1}(p,l,s) + \\
&\quad + p\, e_{3z0t}(p,s) + l\, e_{30zt}(l,s) + e_{300t}(s) = \\
&= \frac{1}{c^2}\left\{\omega_p^2 E_{x1}(p,l,s) + s^2 E_{x1}(p,l,s) - s\, e_{1xz0}(p,l) - e'_{1xz0}(p,l)\right\},
\end{aligned}
$$

$$
\begin{aligned}
&l^2 E_{y1}(p,l,s) - l\, e_{2x0t}(p,s) - e'_{2x0t}(p,s) + \\
&\quad + p^2 E_{y1}(p,l,s) - p\, e_{20zt}(l,s) - e'_{20zt}(l,s) = \\
&= \frac{1}{c^2}\left\{\omega_p^2\left[s E_{y1}(p,l,s) - e_{2xz0}(p,l)\right]\frac{s}{s^2+\omega_b^2} + \right.\\
&\quad \left. + \omega_b \omega_p^2 \frac{s}{s^2+\omega_b^2} E_{z1}(p,l,s) + s^2 E_{y1}(p,l,s) - s\, e_{2xz0}(p,l) - e'_{2xz0}(p,l)\right\},
\end{aligned}
$$

$$
\begin{aligned}
&p^2 E_{z1}(p,l,s) - p\, e_{30zt}(l,s) - e'_{30zt}(l,s) - p l E_{x1}(p,l,s) + \\
&\quad + p\, e_{1x0t}(p,s) + l\, e_{10zt}(l,s) + e_{100t}(s) = \\
&= \frac{1}{c^2}\left\{\omega_p^2 \frac{s}{s^2+\omega_b^2}\left[s E_{z1}(p,l,s) - e_{3xz0}(p,l)\right] - \right.\\
&\quad \left. - \omega_b \omega_p^2 \frac{s}{s^2+\omega_b^2} E_{y1}(p,l,s) + s^2 E_{z1}(p,l,s) - s\, e_{3xz0}(p,l) - e'_{3xz0}(p,l)\right\}.
\end{aligned} \tag{2.154}
$$

As is recognisable in the previous model-calculations, it is not necessary to determine the value of each initial condition by itself, because the coupling equations will deliver a relation for the amplitudes including the initial conditions. Nevertheless, considering the assumed form of the excitations, the

$$
e_i(x,z,0) \equiv 0 \text{ and } e_i'(x,z,0) \equiv 0 \tag{2.155}
$$

suppositions do not mean any restrictions for the solution. Let the following notations be introduced for the remaining initial values

$$\left.\begin{matrix} e_{10zt}(l,s) \\ e_{20zt}(l,s) \\ e_{30zt}(l,s) \end{matrix}\right\} \underset{i=1,2,3}{\longleftrightarrow} A_i(l,s), \qquad \left.\begin{matrix} e_{1x0t}(p,s) \\ e_{2x0t}(p,s) \\ e_{3x0t}(p,s) \end{matrix}\right\} \underset{i=1,2,3}{\longleftrightarrow} B_i(p,s),$$

$$\left.\begin{matrix} e'_{10zt}(l,s) \\ e'_{20zt}(l,s) \\ e'_{30zt}(l,s) \end{matrix}\right\} \underset{i=1,2,3}{\longleftrightarrow} C_i(l,s), \qquad \left.\begin{matrix} e'_{1x0t}(p,s) \\ e'_{2x0t}(p,s) \\ e'_{3x0t}(p,s) \end{matrix}\right\} \underset{i=1,2,3}{\longleftrightarrow} D_i(p,s), \tag{2.156}$$

$$\left.\begin{matrix} e_{100t}(s) \\ e_{300t}(s) \end{matrix}\right\} \underset{i=1,3}{\longleftrightarrow} G_i(s).$$

With (2.156), the transformed solutions are

$$\text{I.} \quad l^2 E_{x1} - lB_1 - D_1 - plE_{z1} + pB_3 + lA_3 + G_3 = \frac{1}{c^2}\left\{\left(\omega_p^2 + s^2\right)E_{x1}\right\},$$

$$\text{II.} \quad l^2 E_{y1} - lB_2 - D_2 + p^2 E_{y1} - pA_2 - C_2 =$$

$$= \frac{1}{c^2}\left\{\left(\frac{\omega_p^2 s^2}{s^2+\omega_b^2} + s^2\right)E_{y1} + \omega_b \omega_p^2 \frac{s}{s^2+\omega_b^2} E_{z1}\right\}, \tag{2.157}$$

$$\text{III.} \quad p^2 E_{z1} - pA_3 - C_3 - plE_{x1} + pB_1 + G_1 =$$

$$= \frac{1}{c^2}\left\{\left(\frac{\omega_p^2 s^2}{s^2+\omega_b^2} + s^2\right)E_{z1} - \omega_b \omega_p^2 \frac{s}{s^2+\omega_b^2} E_{y1}\right\}.$$

After separating the field components, the poles according to p of the denominators may now be determined

$$E_{x1}(p,l,s) = \frac{\left(-c^2 pl\right)E_{z1}(p,l,s)}{\omega_b^2 + s^2 - c^2 l^2} + \frac{c^2\left[pB_3 + lA_3 + G_3 - lB_1 - D_1\right]}{\omega_b^2 + s^2 - c^2 l^2},$$

$$E_{y1}(p,l,s) = \frac{\left(\omega_b \omega_p^2 s\right)E_{z1}(p,l,s)}{\left[c^2\left(l^2+p^2\right) - s^2\right]\left(s^2+\omega_b^2\right) - \omega_p^2 s^2} + \tag{2.158}$$

$$+ \frac{c^2\left[lB_2 + D_2 + pA_2 + C_2\right]}{\left[c^2\left(l^2+p^2\right) - s^2\right]\left(s^2+\omega_b^2\right) - \omega_p^2 s^2}.$$

Although the denominator of E_2 is in function of p to the power 6, the additional virtual two poles will be eliminated – considering that the excitation is z-directed and the coordinate-transformation is indifferent for the y-axis $(A_2 = B_2 = C_2 = D_2 = 0)$. Therefore

$$E_{x1}(p,l,s) = \frac{c^2}{\omega_b^2 + s^2 - c^2 l^2}\left[-pl \cdot E_{z1}(p,l,s) + \mathcal{K}(p,l,s)\right],$$
$$E_{y1}(p,l,s) = \frac{\omega_b \omega_p^2 s}{\left[c^2(l^2+p^2) - s^2\right](s^2+\omega_b^2) - \omega_p^2 s^2} \cdot E_{z1}(p,l,s). \qquad \textbf{(2.159)}$$

The four poles of the field-components, according to p variable are the following

$$p_{1,2}^2 = \pm\frac{NUM}{DEN}$$
$$NUM = \omega_b \omega_p^2 \sqrt{c^4 l^4 \omega_b^2 + 4s^2\left[c^2 l^2 (s^2+\omega_p^2) - (s^2+\omega_p^2)^2\right]} + \qquad \textbf{(2.160)}$$
$$+ 2s^2(s^2+\omega_b^2+\omega_p^2)(s^2+\omega_p^2 - c^2 l^2) + c^2 l^2 \omega_b^2 \omega_p^2$$
$$DEN = 2c^2 \cdot (s^2+\omega_b^2)(s^2+\omega_p^2)$$

Substituting

$$s = \mathrm{j}\omega \text{ and } l = -\mathrm{j}k_z \qquad \textbf{(2.161)}$$

into (2.158), the following forms are given for the poles

$$p_{1,2} = \mp \mathrm{j}k_1 = \mp \mathrm{j}\sqrt{\frac{NUM1 - NUM2}{DEN}},$$

and

$$p_{3,4} = \mp \alpha_3 = \mp\sqrt{\frac{NUM1 + NUM2}{DEN}},$$

where **(2.162)**

$$NUM1 = \omega_b \omega_p^2 \sqrt{c^4 k_z^4 \omega_b^2 + 4\omega^2(\omega_p^2 - \omega^2)\left[c^4 k_z^2 + (\omega_p^2 - \omega^2)\right]},$$
$$NUM2 = \left[2\omega^2(\omega^2 - \omega_b^2 - \omega_p^2)(\omega_p^2 - \omega^2 - c^2 k_z^2) + c^2 k_z^2 \omega_b^2 \omega_p^2\right],$$
$$DEN = 2c^2(\omega^2 - \omega_b^2)(\omega^2 - \omega_p^2).$$

The forms of the field-components are inverse transformable according to p (eliminating the backward propagating terms)

$$\begin{aligned}
E_{x1}(x,\mathrm{j}k_z,\mathrm{j}\omega) &= \frac{E_{0x1}(p_1,\mathrm{j}k_z,\mathrm{j}\omega)}{(p_1-p_2)(p_1-p_3)(p_1-p_4)}\cdot \mathrm{e}^{p_1 x}+ \\
&\quad +\frac{E_{0x1}(p_3,\mathrm{j}k_z,\mathrm{j}\omega)}{(p_3-p_1)(p_3-p_2)(p_3-p_4)}\cdot \mathrm{e}^{p_3 x}, \\
E_{y1}(x,\mathrm{j}k_z,\mathrm{j}\omega) &= \frac{E_{0y1}(p_1,\mathrm{j}k_z,\mathrm{j}\omega)}{(p_1-p_2)(p_1-p_3)(p_1-p_4)}\cdot \mathrm{e}^{p_1 x}+ \\
&\quad +\frac{E_{0y1}(p_3,\mathrm{j}k_z,\mathrm{j}\omega)}{(p_3-p_1)(p_3-p_2)(p_3-p_4)}\cdot \mathrm{e}^{p_3 x}, \\
E_{z1}(x,\mathrm{j}k_z,\mathrm{j}\omega) &= \frac{E_{0z1}(p_1,\mathrm{j}k_z,\mathrm{j}\omega)}{(p_1-p_2)(p_1-p_3)(p_1-p_4)}\cdot \mathrm{e}^{p_1 x}+ \\
&\quad +\frac{E_{0z1}(p_3,\mathrm{j}k_z,\mathrm{j}\omega)}{(p_3-p_1)(p_3-p_2)(p_3-p_4)}\cdot \mathrm{e}^{p_3 x},
\end{aligned} \tag{2.163}$$

and from (2.163)

$$\begin{aligned}
E_{x1}(x,z,t) &= \frac{1}{4\pi^2}\iint_{-\infty}^{\infty} E_{x1}(x,\mathrm{j}k_z,\mathrm{j}\omega)\cdot \mathrm{e}^{-\mathrm{j}k_z z}\cdot \mathrm{e}^{-\mathrm{j}\omega t}\,\mathrm{d}k_z\,\mathrm{d}\omega, \\
E_{y1}(x,z,t) &= \frac{1}{4\pi^2}\iint_{-\infty}^{\infty} E_{y1}(x,\mathrm{j}k_z,\mathrm{j}\omega)\cdot \mathrm{e}^{-\mathrm{j}k_z z}\cdot \mathrm{e}^{-\mathrm{j}\omega t}\,\mathrm{d}k_z\,\mathrm{d}\omega, \\
E_{z1}(x,z,t) &= \frac{1}{4\pi^2}\iint_{-\infty}^{\infty} E_{z1}(x,\mathrm{j}k_z,\mathrm{j}\omega)\cdot \mathrm{e}^{-\mathrm{j}k_z z}\cdot \mathrm{e}^{-\mathrm{j}\omega t}\,\mathrm{d}k_z\,\mathrm{d}\omega.
\end{aligned} \tag{2.164}$$

Further, the transformed magnetic field-components will come from the electric field components, as

$$\begin{aligned}
H_{x1}(p,l,s) &= \frac{l}{\mu_0 s}E_{y1}(p,l,s)-\frac{B_2(p,s)}{\mu_0 s}, \\
H_{y1}(p,l,s) &= \frac{p}{\mu_0 s}E_{z1}(p,l,s)-\frac{A_3(l,s)}{\mu_0 s}-\frac{l}{\mu_0 s}E_{x1}+\frac{B_1(p,s)}{\mu_0 s}, \\
H_{z1}(p,l,s) &= -\frac{p}{\mu_0 s}E_{y1}(p,l,s)+\frac{A_2(l,s)}{\mu_0 s}.
\end{aligned} \tag{2.165}$$

The terms referring solely to the initial conditions are negligible, as the next step in the computation is to solve the coupling equations at the boundary surface, considerations mentioned earlier must be taken into account, and the rotation of the coordinate-system is needed to be executed.

In concordance with Fig.2.48, the original coordinates ("1" indices belonging to I. coordinate system) are determinable in the new coordinate-system ("2" indices – II. coordinate system)

$$\begin{aligned} x_1 &= z_2 \sin\beta + x_2 \cos\beta, \\ y_1 &= y_2, \\ z_1 &= z_2 \cos\beta - x_2 \sin\beta. \end{aligned} \tag{2.166}$$

As is obvious, medium "1" has not been modified in the present case in comparison with the previous one, so the solution presented in part 1.3.4 will be valid now as well. All the possibly existing modes have to be taken into consideration in the coupling equations.

In medium "1" the modes are

$$\begin{aligned} E_{1y_2R}(x,t) &= \frac{1}{2\pi}\int_{-\infty}^{\infty} E_{1y_2R}(\omega)\mathrm{e}^{j\omega\left(t+\frac{x}{c}\right)}\mathrm{d}\omega, \\ E_{1z_2G}(x,t) &= \frac{-Z_0}{2}\int_{0}^{x_0} J_0\left(\xi, t-\frac{x-\xi}{c}\right)\mathrm{d}\xi, \\ E_{1z_2R}(x,t) &= \frac{1}{2\pi}\int_{-\infty}^{\infty} E_{1z_2R}(\omega)\mathrm{e}^{j\omega\left(t+\frac{x}{c}\right)}\mathrm{d}\omega, \\ H_{1y_2R}(x,t) &= \frac{1}{Z_0} E_{1z_2R}(x,t), \\ H_{1z_2R}(x,t) &= -\frac{1}{Z_0} E_{1y_2R}(x,t), \\ H_{1y_2G}(x,t) &= -\frac{1}{Z_0} E_{1z_2G}(x,t). \end{aligned} \tag{2.167}$$

where the first index – 1 – refers to medium "1", the y and z denote the components, the second index-number – 2 – indicates the coordinate-system II., the G and R indices represent the characters of the modes.

The coupling equations delivered by the application of MIBM for Maxwell's equations are the following (the first number 1 and 2 refer to medium "1" and "2" in the indices x, y and z are the components, the second number means the coordinate-system II., w – whistler-mode; N – non propagating; R – reflected; G – generated directly by excitation)

$$\begin{aligned}
&\text{I.}\quad H_{1z_2\mathrm{R}}(x_0,t)=H_{2z_2\mathrm{w}}(x_0,z,t)+H_{2z_2\mathrm{N}}(x_0,z,t),\\
&\text{II.}\quad H_{1y_2\mathrm{G}}(x_0,t)+H_{1y_2\mathrm{R}}(x_0,t)=H_{2y_2\mathrm{w}}(x_0,z,t)+H_{2y_2\mathrm{N}}(x_0,z,t),\\
&\text{III.}\quad E_{1z_2\mathrm{G}}(x_0,t)+E_{1z_2\mathrm{R}}(x_0,t)=E_{2z_2\mathrm{w}}(x_0,z,t)+E_{2z_2\mathrm{N}}(x_0,z,t),\\
&\text{IV.}\quad E_{1y_2\mathrm{R}}(x_0,t)=E_{2y_2\mathrm{w}}(x_0,z,t)+E_{2y_2\mathrm{N}}(x_0,z,t),\\
&\text{V.}\quad 0=H_{2x_2\mathrm{w}}(x_0,z,t)+H_{2x_2\mathrm{N}}(x_0,z,t),\\
&\text{VI.}\quad 0=E_{2x_2\mathrm{w}}(x_0,z,t)+E_{2x_2\mathrm{N}}(x_0,z,t).
\end{aligned}\tag{2.168}$$

As can be seen in (2.167), the left side of the equations does not depend on z, therefore the z-dependence has to disappear from the right side (at $x=x_0$). Further, it must be kept in mind that the propagation factors (k_x and k_z) were determined in coordinate system I. in the medium "2", so this fact has to be indicated by their indices in the transformed coordinate system II. (k_{x1}, k_{z1}). In equation IV

$$\begin{aligned}
E_{1y_2\mathrm{R}}(x_0 t)=\frac{1}{4\pi^2}\Bigg[&\iint\limits_{-\infty}^{\infty} E_{2y_1\mathrm{w}}\cdot \mathrm{e}^{\mathrm{j}[\omega t-(k_{x1}\sin\beta+k_{z1}\cos\beta)z_2]}\,\mathrm{d}k_{z1}\,\mathrm{d}\omega+\\
&+\iint\limits_{-\infty}^{\infty} E_{2y_1\mathrm{N}}\cdot \mathrm{e}^{\mathrm{j}[\omega t-k_{z1}z_2\cos\beta]}\cdot \mathrm{e}^{-\alpha_{x1}z_2\cos\beta}\,\mathrm{d}k_{z1}\,\mathrm{d}\omega\Bigg],
\end{aligned}\tag{2.169}$$

where

$$-\mathrm{j}\left[k_{x_1}\sin\beta+k_{z_1}\cos\beta\right]z_2\equiv 0 \tag{2.170}$$

and

$$-\left[\mathrm{j}k_{z_1}\cos\beta+\alpha_{x_1}\sin\beta\right]z_2\equiv 0. \tag{2.171}$$

k_{z1} can be derived from (2.170) and (2.171)

$$
\begin{aligned}
& \mathrm{j}k_{z_1\mathrm{p}} \triangleq \mathrm{j}k_{z_1} = \left(-\mathrm{j}k_{x_1}\right)\mathrm{tg}\,\beta \ , \\
& \alpha_{z_1\mathrm{p}} \triangleq \mathrm{j}k_{z_1} = \left(-\alpha_{x_1}\right)\mathrm{tg}\,\beta \, , \\
& l = -p\,\mathrm{tg}\,\beta \, ,
\end{aligned}
\tag{2.172}
$$

in concordance with the refraction-reflection law.

Substituting $l=-p\,\mathrm{tg}\,\beta$ into $E_{2z_1}(p,l,s)$ the poles depending on the frequency and angle of propagation will be given ($s=\mathrm{j}\omega$)

$$
p_1 = -\mathrm{j}k_{x1} =
$$

$$
= \mathrm{j}\frac{\cos\beta}{\sqrt{2}c}\sqrt{\frac{\omega\omega_b\omega_p^2\sqrt{4\left(\omega^2-\omega_p^2\right)^2\cos^2\beta+\omega^2\omega_b^2\sin^4\beta}+\omega^2\left[\omega_b^2\omega_p^2\cos^2\beta+2\omega^4+\left(\omega_p^2-2\omega^2\right)\left(\omega_b^2+2\omega_p^2\right)\right]}{\omega_b^2\omega_p^2\cos^2\beta+\omega^2\left(\omega^2-\omega_b^2-\omega_p^2\right)}} \, ,
$$

$$
p_3 = -\alpha_{x1} =
$$

$$
= \frac{\cos\beta}{\sqrt{2}c}\sqrt{\frac{\omega\omega_b\omega_p^2\sqrt{4\left(\omega^2-\omega_p^2\right)^2\cos^2\beta+\omega^2\omega_b^2\sin^4\beta}-\omega^2\left[\omega_b^2\omega_p^2\cos^2\beta+2\omega^4+\left(\omega_p^2-2\omega^2\right)\left(\omega_b^2+2\omega_p^2\right)\right]}{\omega_b^2\omega_p^2\cos^2\beta+\omega^2\left(\omega^2-\omega_b^2-\omega_p^2\right)}} \, .
\tag{2.173}
$$

Writing back (2.171) into the modes existing in medium"2", the form of e.g. $E_{2z_1\mathrm{w}}$ is the following

$$
\begin{aligned}
E_{2z_1\mathrm{w}}(x,z,t) &= \frac{1}{4\pi^2}\iint\limits_{-\infty}^{\infty} E_{2z_1\mathrm{w}}(\omega,\beta)2\pi\,\delta\!\left(k_{z_1}-k_{z_1\mathrm{p}}\right)\cdot \mathrm{e}^{\mathrm{j}\left(\omega t-k_{x_1}x_1-k_{z_1}z_1\right)}\,\mathrm{d}k_{z_1}\,\mathrm{d}\omega = \\
&= \frac{1}{2\pi}\int\limits_{-\infty}^{\infty} E_{2z_1\mathrm{w}}(\omega,\beta)\cdot \mathrm{e}^{\mathrm{j}\left(\omega t-k_{x_1}x_1-k_{z_1\mathrm{p}}z_1\right)}\,\mathrm{d}\omega .
\end{aligned}
\tag{2.174}
$$

The modes of medium "2" in the new coordinate-system are the following

$$E_{2x_2\mathrm{w}}(x,z,t)=\frac{1}{2\pi}\int_{-\infty}^{\infty}\left[E_{2x_1\mathrm{w}}\cos\beta-E_{2z_1\mathrm{w}}\sin\beta\right]\cdot \mathrm{e}^{\mathrm{j}\left[\omega t-\left(k_{x_1}\cos\beta-k_{z_1\mathrm{P}}\sin\beta\right)(x_2-x_{20})-\underbrace{\left(k_{x_1}\sin\beta+k_{z_1\mathrm{P}}\cos\beta\right)z_2}_{\equiv 0}\right]}\mathrm{d}\omega,$$

$$E_{2x_2\mathrm{N}}(x,z,t)=\frac{1}{2\pi}\int_{-\infty}^{\infty}\left[E_{2x_1\mathrm{N}}\cos\beta-E_{2z_1\mathrm{N}}\sin\beta\right]\cdot \mathrm{e}^{\mathrm{j}\omega t-\left(\alpha_{x_1}\cos\beta-\alpha_{z_1\mathrm{P}}\sin\beta\right)(x_2-x_{20})-\underbrace{\left(\alpha_{x_1}\sin\beta+\alpha_{z_1\mathrm{P}}\cos\beta\right)z_2}_{\equiv 0}}\mathrm{d}\omega,$$

$$E_{2y_2\mathrm{w}}(x,z,t)=\frac{1}{2\pi}\int_{-\infty}^{\infty}E_{2y_1\mathrm{w}}\cdot \mathrm{e}^{\mathrm{j}\left[\omega t-\left(k_{x_1}\cos\beta-k_{z_1\mathrm{P}}\sin\beta\right)(x_2-x_{20})-\underbrace{\left(k_{x_1}\sin\beta+k_{z_1\mathrm{P}}\cos\beta\right)z_2}_{\equiv 0}\right]}\mathrm{d}\omega,$$

$$E_{2y_2\mathrm{N}}(x,z,t)=\frac{1}{2\pi}\int_{-\infty}^{\infty}E_{2y_1\mathrm{N}}\cdot \mathrm{e}^{\mathrm{j}\omega t-\left(\alpha_{x_1}\cos\beta-\alpha_{z_1\mathrm{P}}\sin\beta\right)(x_2-x_{20})-\underbrace{\left(\alpha_{x_1}\sin\beta+\alpha_{z_1\mathrm{P}}\cos\beta\right)z_2}_{\equiv 0}}\mathrm{d}\omega, \tag{2.175a}$$

$$E_{2z_2\mathrm{w}}(x,z,t)=\frac{1}{2\pi}\int_{-\infty}^{\infty}\left[E_{2x_1\mathrm{w}}\sin\beta+E_{2z_1\mathrm{w}}\cos\beta\right]\cdot \mathrm{e}^{\mathrm{j}\left[\omega t-\left(k_{x_1}\cos\beta-k_{z_1\mathrm{P}}\sin\beta\right)(x_2-x_{20})-\underbrace{\left(k_{x_1}\sin\beta+k_{z_1\mathrm{P}}\cos\beta\right)z_2}_{\equiv 0}\right]}\mathrm{d}\omega,$$

$$E_{2z_2\mathrm{N}}(x,z,t)=\frac{1}{2\pi}\int_{-\infty}^{\infty}\left[E_{2x_1\mathrm{N}}\sin\beta+E_{2z_1\mathrm{N}}\cos\beta\right]\cdot \mathrm{e}^{\mathrm{j}\omega t-\left(\alpha_{x_1}\cos\beta-\alpha_{z_1\mathrm{P}}\sin\beta\right)(x_2-x_{20})-\underbrace{\left(\alpha_{x_1}\sin\beta+\alpha_{z_1\mathrm{P}}\cos\beta\right)z_2}_{\equiv 0}}\mathrm{d}\omega,$$

$$H_{2x_2\mathrm{w}}(x,z,t)=0,$$

$$H_{2x_2\mathrm{N}}(x,z,t)=0,$$

$$H_{2y_2\mathrm{w}}(x,z,t)=\frac{1}{2\pi}\int_{-\infty}^{\infty}\left[-\frac{k_{x_1}}{k_0\,z_0}E_{2z_1\mathrm{w}}+\frac{k_{z_1\mathrm{P}}}{k_0\,z_0}E_{2x_1\mathrm{w}}\right]\cdot \mathrm{e}^{\mathrm{j}\left[\omega t-\left(k_{x_1}\cos\beta-k_{z_1\mathrm{P}}\sin\beta\right)(x_2-x_{20})-\underbrace{\left(k_{x_1}\sin\beta+k_{z_1\mathrm{P}}\cos\beta\right)z_2}_{=0}\right]}\mathrm{d}\omega, \qquad \textbf{(2.175b)}$$

$$H_{2y_2\mathrm{N}}(x,z,t)=\frac{1}{2\pi}\int_{-\infty}^{\infty}\left[\mathrm{j}\frac{\alpha_{x_1}}{k_0\,z_0}E_{2z_1\mathrm{N}}-\mathrm{j}\frac{\alpha_{z_1\mathrm{P}}}{k_0\,z_0}E_{2x_1\mathrm{N}}\right]\cdot \mathrm{e}^{\mathrm{j}\omega t-\left(\alpha_{x_1}\cos\beta-\alpha_{z_1\mathrm{P}}\sin\beta\right)(x_2-x_{20})-\underbrace{\left(\alpha_{x_1}\sin\beta+\alpha_{z_1\mathrm{P}}\cos\beta\right)z_2}_{=0}}\mathrm{d}\omega,$$

$$H_{2z_2\mathrm{w}}(x,z,t)=\frac{1}{2\pi}\int_{-\infty}^{\infty}\left[\frac{k_{x_1}}{k_0\,z_0}\frac{1}{\cos\beta}E_{2y_1\mathrm{w}}\right]\cdot \mathrm{e}^{\mathrm{j}\left[\omega t-\left(k_{x_1}\cos\beta-k_{z_1\mathrm{P}}\sin\beta\right)(x_2-x_{20})-\underbrace{\left(k_{x_1}\sin\beta+k_{z_1\mathrm{P}}\cos\beta\right)z_2}_{=0}\right]}\mathrm{d}\omega,$$

$$H_{2z_2\mathrm{N}}(x,z,t)=\frac{1}{2\pi}\int_{-\infty}^{\infty}\left[-\mathrm{j}\frac{1}{\cos\beta}\frac{\alpha_{x_1}}{k_0\,z_0}E_{2y_1\mathrm{N}}\right]\cdot \mathrm{e}^{\mathrm{j}\omega t-\left(\alpha_{x_1}\cos\beta-\alpha_{z_1\mathrm{P}}\sin\beta\right)(x_2-x_{20})-\underbrace{\left(\alpha_{x_1}\sin\beta+\alpha_{z_1\mathrm{P}}\cos\beta\right)z_2}_{=0}}\mathrm{d}\omega,$$

where the amplitudes of the field-components are obtained from (2.159)

$$E_{2x_1w}(\omega)=\frac{c^2}{\omega_p^2-\omega^2+c^2k_{x1}^2\,\mathrm{tg}^2\beta}\left[-k_{x1}^2\,\mathrm{tg}\beta+\frac{\mathcal{K}(k_{x1},\omega)}{E_{2z_1w}(\omega)}\right]E_{2z_1w}(\omega)=M_{xw}(\omega)\cdot E_{2z_1w}(\omega),$$

$$E_{2x_1N}(\omega)=\frac{c^2}{\omega_p^2-\omega^2-c^2\alpha_{x1}^2\,\mathrm{tg}^2\beta}\left[\alpha_{x1}^2\,\mathrm{tg}\beta+\frac{\mathcal{K}(\alpha_{x1},\omega)}{E_{2z_1N}(\omega)}\right]E_{2z_1N}(\omega)=M_{xN}(\omega)\cdot E_{2z_1N}(\omega),$$

$$E_{2y_1w}(\omega)=E_{2z_1w}(\omega)\cdot\frac{\mathrm{j}\omega\omega_b\omega_p^2}{\omega^2\omega_p^2+(\omega_b^2-\omega^2)[\omega^2-c^2k_{x1}^2(1+\mathrm{tg}^2\beta)]}=M_{yw}(\omega)\cdot E_{2z_1w}(\omega),$$

$$E_{2y_1N}(\omega)=E_{2z_1N}(\omega)\cdot\frac{\mathrm{j}\omega\omega_b\omega_p^2}{\omega^2\omega_p^2+(\omega_b^2-\omega^2)[\omega^2+c^2\alpha_{x1}^2(1+\mathrm{tg}^2\beta)]}=M_{yN}(\omega)\cdot E_{2z_1N}(\omega).$$

(2.176)

Writing (2.175) and (2.176) into (2.168) the following equations will be obtained for the amplitude-relations

from I. + IV.:

$$E_{2y_1w}(k_0\cos\beta+k_{x1})=E_{2y_1N}(-k_0\cos\beta+\mathrm{j}\alpha_{x1}),$$
$$E_{2z_1w}\cdot M_{yw}(k_0\cos\beta+k_{x1})=E_{2z_1N}\cdot M_{yN}(-k_0\cos\beta+\mathrm{j}\alpha_{x1}); \qquad \textbf{(2.177)}$$

from VI:

$$E_{2z_1w}(M_{xw}\cos\beta-\sin\beta)=E_{2z_1N}(-M_{xN}\cos\beta+\sin\beta). \qquad \textbf{(2.178)}$$

(2.177) and (2.178) deliver a prescription for initial values, as

$$-M_{xN}\cos\beta+\sin\beta\triangleq M_{yN}\left(-\cos\beta+\mathrm{j}\frac{\alpha_{x1}}{k_0}\right),$$
$$M_{xw}\cos\beta-\sin\beta\triangleq M_{yw}\left(\cos\beta+\frac{k_{x1}}{k_0}\right), \qquad \textbf{(2.179)}$$

where M_{xw} and M_{xN} contain the additional term $\mathcal{K}$ including initial conditions. Although this term seems at first to be independent from the complete propagational field, the elimination of that would cause contradiction in the coupling equations. The physical meaning of this term is actually not independent of the propagating signal, but describes a "cross-section" of the field at a given direction ($z_1=0$ and $x_1=0$), in which the amplitude-relations are fixed. An essential difference between the present and the previous ones is the space-dependence of the initial values ("initial planes").

The connection between the excitation and the signal of medium "2" will be obtained from III. – V. (with I.+IV):

$$
\begin{aligned}
E_{2z_1\mathrm{w}}(\omega) &= \frac{NUM}{DEN} \\
NUM &= -Z_0\, I_{x0} \\
DEN &= \frac{k_{x1}+k_0\cos\beta}{k_0\cos\beta}\left\{\sin\beta\left[M_{y\mathrm{w}}\frac{k_0\cos\beta+k_{x1}}{k_0\cos\beta}+\mathrm{tg}\beta\right]-\right. \\
&\left.-\sin\beta\left[M_{y\mathrm{w}}\frac{k_0\cos\beta-\mathrm{j}\alpha_{x1}}{k_0\cos\beta}+\frac{M_{y\mathrm{w}}}{M_{y\mathrm{N}}}\mathrm{tg}\,\beta\right]+1-\frac{M_{y\mathrm{w}}}{M_{y\mathrm{N}}}\right\},
\end{aligned}
\qquad \mathbf{(2.180)}
$$

where $k_0=\omega/c$ and I_{x0} is given by equation (1.67). The accurate form of the signal can be obtained from (2.175) using (2.176) and (2.180)

$$
E_{2z\mathrm{w}}(x,t)=\frac{1}{2\pi}\int_{-\infty}^{\infty}\left[E_{2x_1\mathrm{w}}(\omega)\sin\beta+E_{2z_1\mathrm{w}}(\omega)\cos\beta\right]\cdot \mathrm{e}^{\mathrm{j}\left[\omega t-k_{x1}\frac{1}{\cos\beta}(x_2-x_{20})\right]}\mathrm{d}\omega. \qquad \mathbf{(2.181)}
$$

2.5.4 Numerical results

In the calculation of the results presented here, a Dirac-delta excitation and homogeneous plasma were considered.

Fig. 2.49 shows a resultant interference in the FFT-pattern of whistlers propagating to oblique directions with small deviation-angles to the geomagnetic field line. The influence of this effect is a fused spectrum with random maxima and minima, similar to the spectrum of a single whistler often experienced in the observations.

The FFT-pattern for eight different angles can be seen in Fig. 2.50. These calculated FFT-patterns are very similar to a noisy, “swishy” signal well known from the measurements, which is often interpreted by the assumption of the propagation in near-by ducts.

The FFT-patterns in the case of large deviation-angles are shown in Fig. 2.51. Similar phenomena are often detected and interpreted as multi-path propagation effects, or assumed to be caused by multiple exciting events.

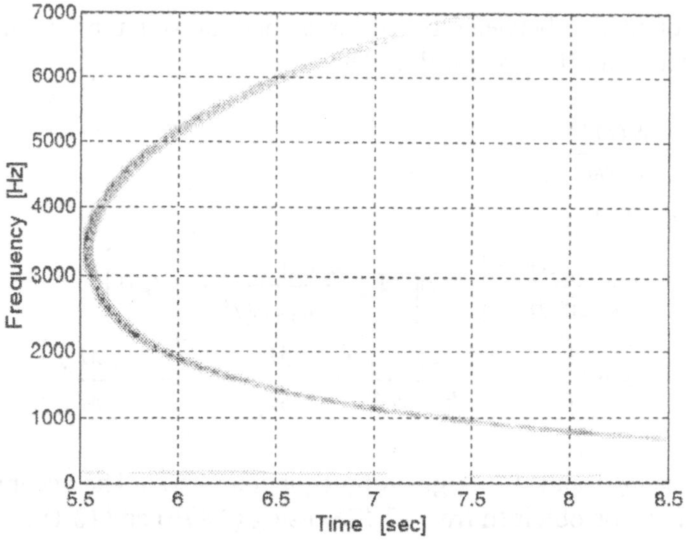

Figure 2.49. **FFT-pattern of whistlers for oblique propagation (β=0°, 1°, 2°, 3°, 4°, 5°, 6°)**

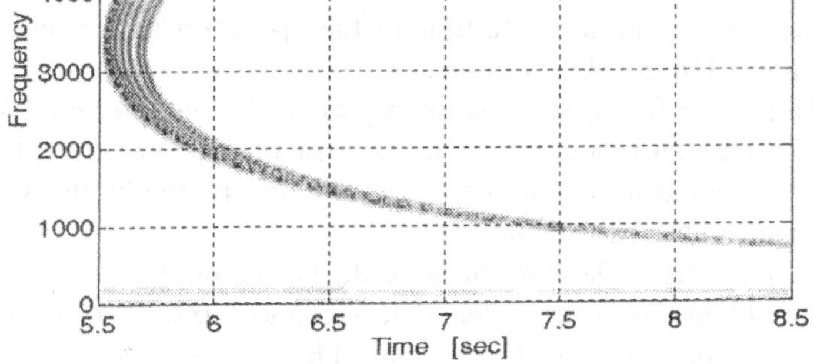

Figure 2.50. **FFT-pattern of whistlers for oblique propagation (β=0°, 3°, 5°, 7°, 9°, 11°, 13°, 14°)**

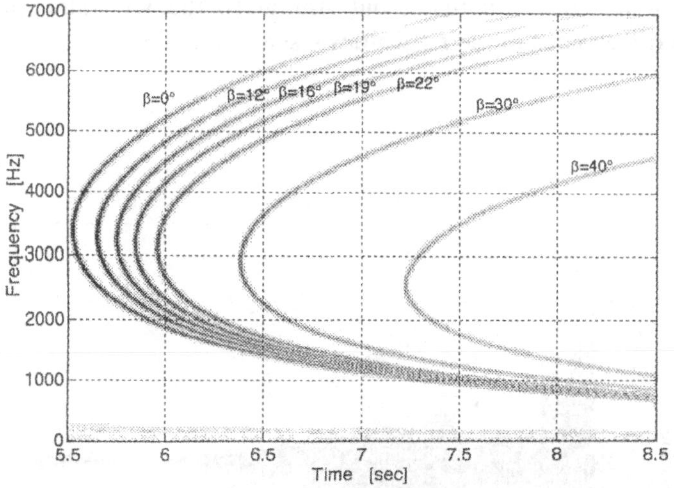

Figure 2.51. FFT-pattern of whistlers for oblique propagation (β=0°, 12°, 16°, 19°, 22°, 30°, 40°)

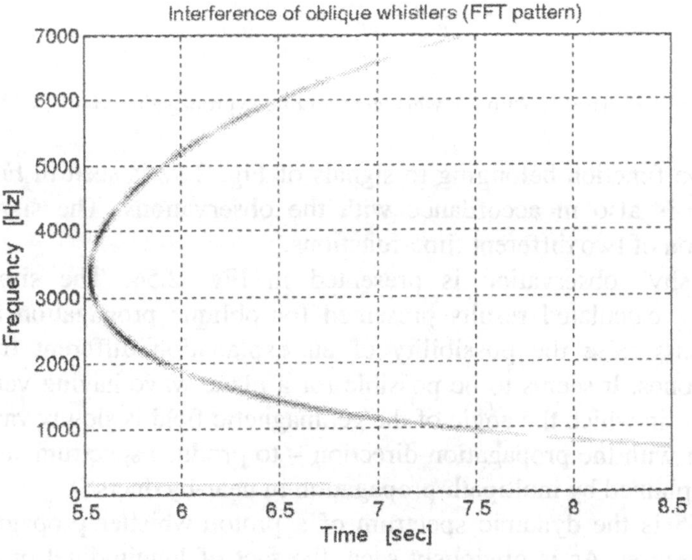

Figure 2.52. FFT-pattern of whistlers for oblique propagation (β=0°, 1.2°)

The interference of dynamic spectra of very small angles is shown in Fig. 2.52. The superposition of two whistlers, travelling on the same path, forms one signal. The positions of the minima are not periodical (contrary to the influence of a square impulse), but similar to the whistlers travelling on paths with differences in their L-values – see Table 4.3.

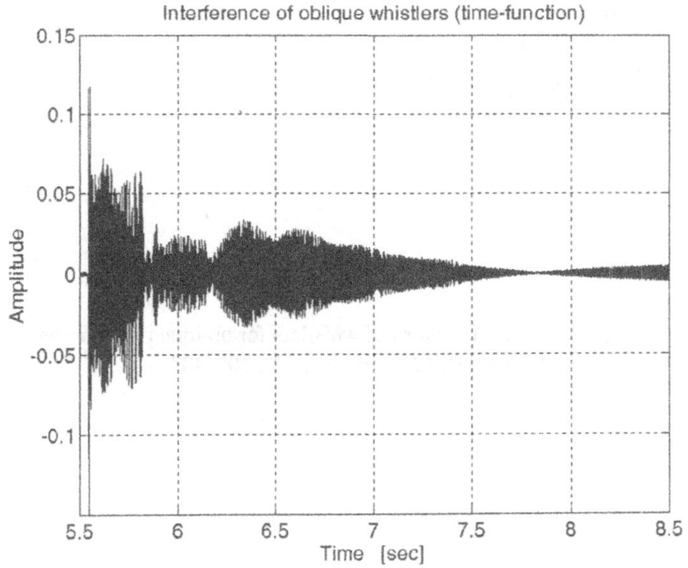

Figure 2.53. Time function of whistlers for oblique propagation (β=0°, 1.2°)

The time-function belonging to signals of Fig. 2.52 is seen in Fig. 2.53. This result is also in accordance with the observations. The signal is a superposition of two different time-functions.

A "swishy" observation is presented in Fig. 2.54. The similarities between the calculated results presented for oblique propagation and the measurements raise the possibility of an explanation different from the traditional ones. It seems to be possible for a plane-wave having very large wave front – in which the angle of the geomagnetic field is slowly varying in comparison with the propagation-direction – to produce spectrum similar to the ones explained by multipath-propagation in near-by ducts.

Fig. 2.55 is the dynamic spectrum of a proton-whistler propagating to oblique direction. As is obviously seen, the fact of longitudinal or oblique propagation does not influence or change the number and the character of the existing modes, in concordance with the results presented in earlier parts.

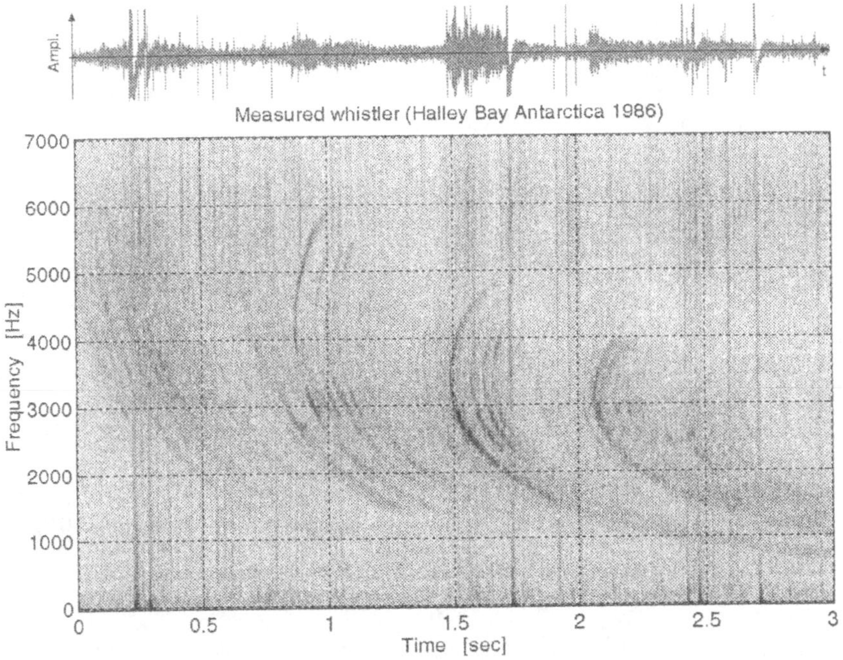

Figure 2.54. "Swishy" whistler observation (No. TS86147 recorded at Halley, Antarctica, 10 August 1986 14:52:09 UT, courtesy of A.J. Smith, British Antarctic Survey, Cambridge, U.K.)

A cornerstone of the traditional whistler-theory is the assumption of the polarization-reversal in order to explain some observed effect (e.g. the generation of ion-whistlers). In these approaches, the oblique propagation – where the direction of propagation deviates from the geomagnetic field line – is indispensable for the explanation of the registrations (*Rodriguez and Gurnett*, 1971; *Wang*, 1971; *Fijalkow et al.*, 1973; *Walker*, 1993). Contrary to these hypotheses, the rotation of the signal can be continuously analysed knowing the exact time-space dependence obtained from the new full-wave model. Fig. 2.56 shows the curve drawn by the end-point of the resultant field-intensity vector in the case of an obliquely propagating whistler. (The influence of the E_x component is marginal only, causing an oscillation of the plane of elliptical polarization through the $\pm\ x$ directions, but this effect cannot modify the character of rotation. The maximum of this effect is 90° in the case of transversal propagation.) As can be seen, the direction of the (elliptical) rotation of the signal is unchanged, and no polarization reversal happens.

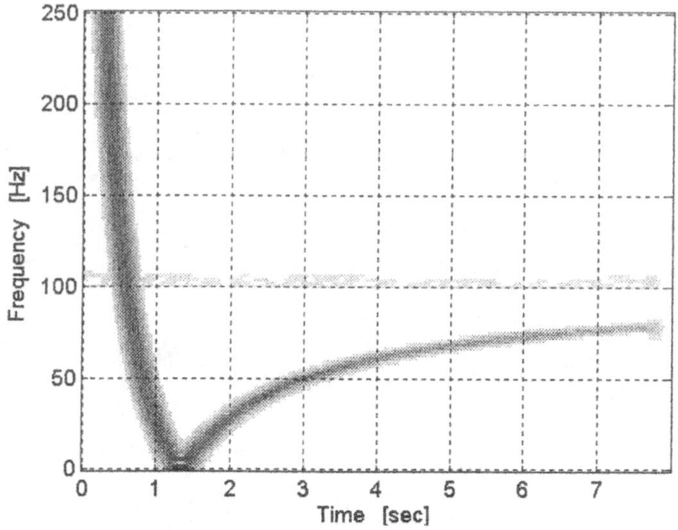

Figure 2.55. FFT-pattern of an obliquely propagating proton-whistler ($\beta=10°$)

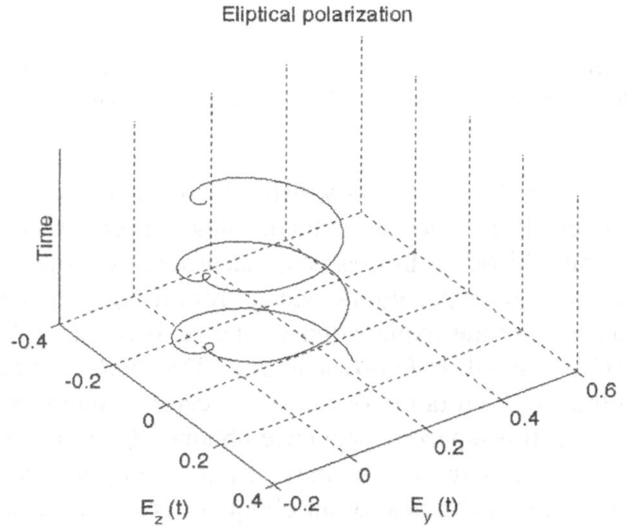

Figure 2.56. Polarisation of an obliquely propagating whistler

Table 2.2 contains the result of the parameter-estimation of the FIT-algorithm. The propagation paths are the same in each case, but the angles between the geomagnetic field and the direction of the propagation are different. The variation of the estimated values strongly depends on the angle. The source of this error is hidden in the FIT-algorithm that evaluate every signal as if the direction of propagation was longitudinal. Consequently, the signal propagating obliquely is assumed to be longitudinal, and the estimated parameters definitely differ from the actual, real values. The longitudinal and the non-longitudinal propagation are not separable perfectly and unambiguously. In whistler investigations, many non-longitudinal, oblique whistlers were probably analysed as if they were longitudinally propagating ducted whistlers (e.g. *L*-discrepancy).

Table 2.2 Estimated parameters based on the Eckersley's model

β [°]	L	D [$s^{-0.5}$]	N_{1000} [cm^{-3}]
0	5.266	48.2	484
1	5.266	48.4	488
5	5.272	48.9	496
10	5.293	49.0	499

2.5.5 Conclusions

It is possible to apply the theoretical model for oblique propagation with essential modifications. The possibilities and limits of its application can be seen in part 2.5.2.

It has become clear that no essential difference appears in the time-depending form (time functions of calculated whistlers) and the dynamic spectrum of the signal in comparison with the longitudinally propagating ones, except the decreasing nose-frequency and amplitude and increasing time delays, depending on the increase of the angle to the magnetic field. It is undoubtedly proved by the calculations that no polarisation reversal happens in the case of obliquely propagating whistlers, the same independent (EM and HM) modes exist in this case as in the longitudinal case.

Therefore, the assumption of unchanged forms of whistlers scattered out and detected by satellite – see *Cerisier's* approach (*Tarcsai*, 1981) or others (eg. *Thomson* and *Dowden*, 1977a,b) – in a dispersive medium must be false (in accordance with the wave-theory).

As a consequence of the (homogeneous) model calculations, the reconsideration of the explanations of generation-mechanism of multiple-traces (*Helliwell*, 1965; *Walker*, 1993) seems to be necessary, as an extended VLF wave-front, in which the angle between the direction of the magnetic field and the propagation-vector is varying continuously, may produce similar dynamic spectra to that of the observations. This similarity makes a new explanation of the *L*-discrepancy likely, contrary to the scattering of whistlers from ducts above the satellite. Moreover, the assumption of the existence of near-by ducts is not the one and only possible explanation of the measured data.

As a first approximation of an obliquely propagating proton-whistler (its exact amplitude-variation has been missing up to now) shows the dynamic spectrum, there no theoretical singularity appears between the longitudinal and oblique propagation, confirming indirectly the conclusion, that the oblique propagation is not a precondition of generation of proton- (ion-) whistlers.

Further, it is verified by the results, that neither the longitudinal, nor the other cases of the propagation are anomalous – in opposition to *Walker* (1993), but in concordance with the wave-propagation theory and the consistent picture of the physical world. Ion-whistlers may occur in the case of longitudinal propagation as well as of oblique one, without assuming any kind of polarisation-reversal or mode-coupling. The fact that the estimated direction of the whistlers registered by satellites is oblique, may be originated from the same effect as the *L*-discrepancy, however, this cannot preclude the possibility of the longitudinal propagation.

Chapter 3

MEASURING OF GENERAL SHAPE ELECTROMAGNETIC SIGNALS OF NATURAL ENVIRONMENT

In this chapter we will consider ELF-VLF waves because of the following reasons:

- The speed of the generally available electronic components, namely the commercially available analogue-to-digital converters, the computation speed of the computers and in space experiments the effective capacity of the typical telemetry-systems prevents us from dealing with high resolution detailed time-domain analyses of higher frequency e.m. waves in the routine monitoring of electromagnetic environment.
- In the Earth's upper atmosphere the propagation factor of these low frequency waves makes the analyses possible.
- All the results obtained from the analyses of ELF-VLF waves are valid to other higher frequency signals propagating in similar media (see point 2.4.3).

However, the advances in R+D will produce new technical opportunities in the near future and therefore the detailed broad-band high-resolution e.m. signal detection and analyses will appear in higher frequency bands too, as routine monitoring techniques.

3.1 Ground based measurements

In the case of ground based measurements, instead of measuring six components of the e.m. waves, the two horizontal magnetic and the vertical electric components can be measured because of the conductivity of the surface.

Though the Earth's surface is not a perfect conductor, the ELF-VLF waves propagate in the Earth-ionosphere waveguide for long distances – several hundreds or even thousands of km –, during this propagation part of the wave energy is dissipated: the horizontal electric and the vertical magnetic components generate (surface) currents because the conductivity is not zero. This dissipation happens faster (i.e. within shorter distance) above water and snow or ice covered land then over normal land surface, simply because of the higher conductivity of water and ice. Also, the Earth's currents would add high level noise to these components – however, in the case of Earth-related measurements, just above the volume of interest under the ground, these components hold relevant information (earthquake-related investigations, mine-search etc.)

In the case of wave propagation related measurement (e.g. investigation of polarisation, of wave normal) the ground based measurements have serious limitations partly because of the limited number of the measured e.m. field components and therefore the investigated parameters cannot be derived exactly for elliptically polarized waves. Also, a wave exiting from the bottom of ionosphere into the higher atmosphere may propagate to a distant receiver more than one path, i.e. reflecting one or more times between the surface and the ionosphere, the resulted wave front is the interference of these waves or a guided wave pattern of a single mode or multi mode propagation and yields erroneous result in propagation parameters (see part 4.3 for more detail). Another problem is that the upgoing signals that just start to propagate through the magnetosphere evidently cannot be examined by ground based measurements – this is the area where satellite measurements are unique. In higher frequency bands, the signals have a general upgoing character and they can be investigated also by satellites only.

In ground based measurements the horizontal magnetic field components are measured by two orthogonal crossed-loop antennas, usually with one or a few turns and with an area as large as possible (several tens or hundred m^2) to optimise the signal-to-noise ratio. The electric field components are measured by the voltage difference between a sphere and the ground. Up to the 1980's, the recordings were made on analogue tapes, at present, the signals are directly digitised and stored. The ELF-VLF measurements are subject to the noise of human activity (power lines, transportation etc.) thus, the receivers are set up in electromagnetically quiet places – outside cities or in uninhabited areas like the Antarctica. Detailed description of such receivers see e.g. *Bullough and Sagredo* (1973) and *Smith*, (1995). The ELF-VLF recordings are used to investigate the amplitude, the phase, the polarization, the frequency-time variation and the direction of the arrival of the signal. These characteristic are examined by the analyses of different natural signals (whistlers, choruses, hisses, bursts, emissions, – see examples in Fig. 3.1) propagating through the ionosphere and magnetosphere and by the analyses of the signal of VLF transmitters that are also connected to natural phenomena (trimpi effect and related phenomena, – see example in Fig. 3.2).

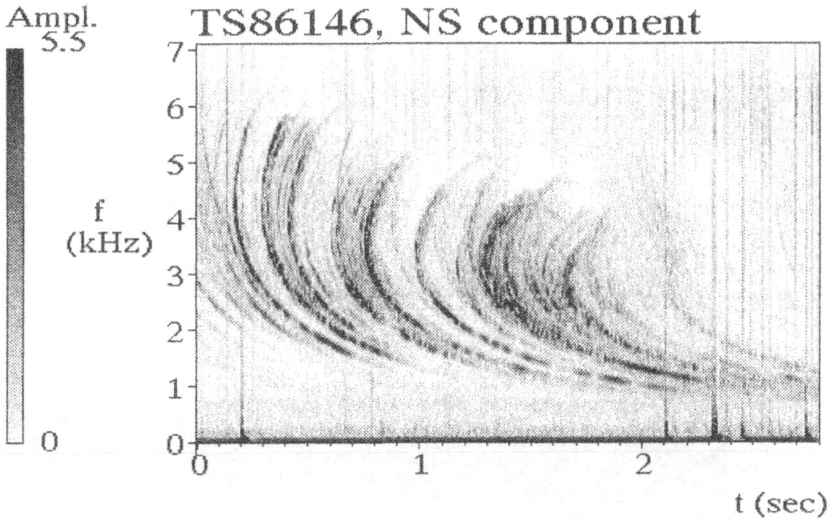

Figure 3.1a. High latitude whistler groups recorded at Halley, Antarctica on 10 August 1986, 14:48:13 UT. (Courtesy of dr. A.J. Smith, British Antarctic Survey, Cambridge, U.K.)

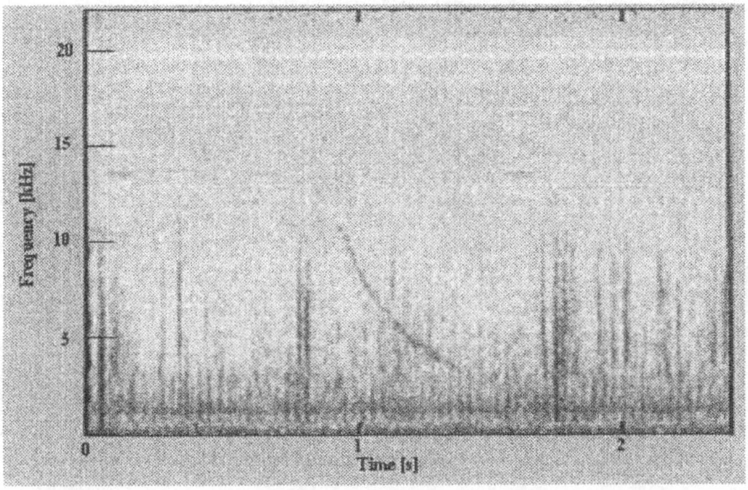

Figure 3.1b. Medium-latitude whistler recorded in Tihany, Hungary on 4 March 1970, 20:54:44 UT.

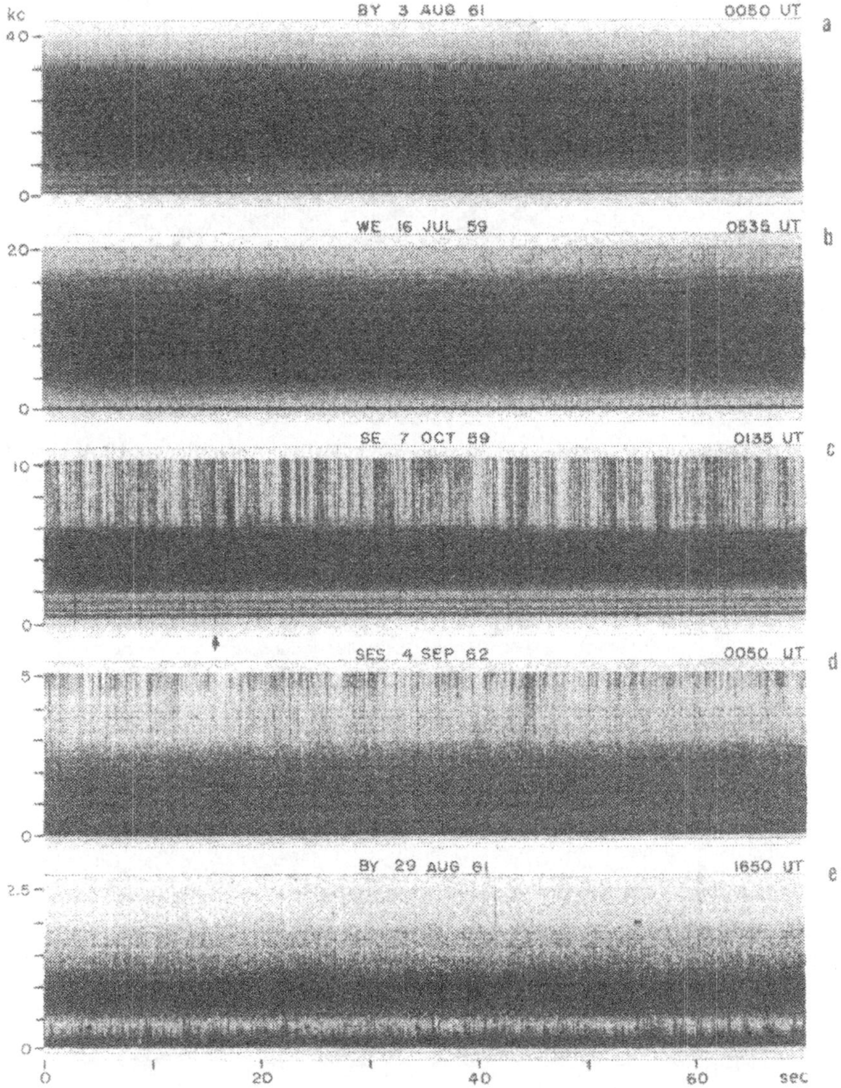

Figure 3.1c. VLF hisses with varying upper cut-off frequency.
(From Helliwell, 1965, Fig.7-1.)

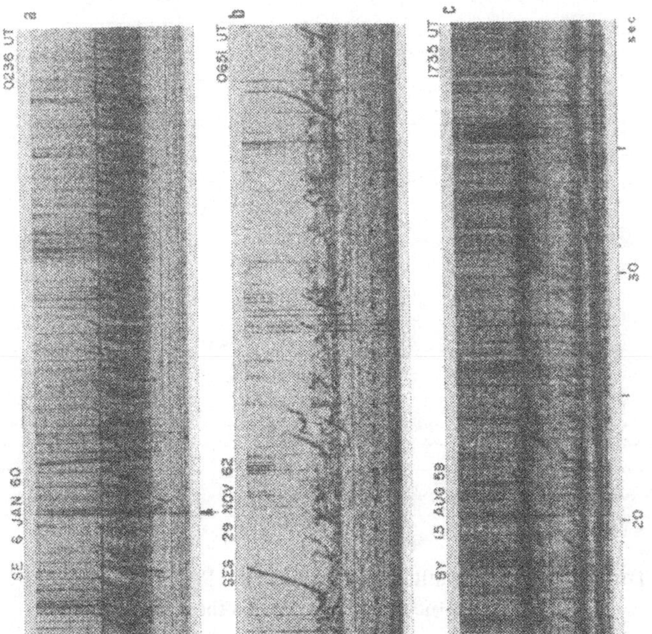

Figure 3.1d VLF choruses, emissions and risers. (From Helliwell, 1965, - Fig.7-34.)

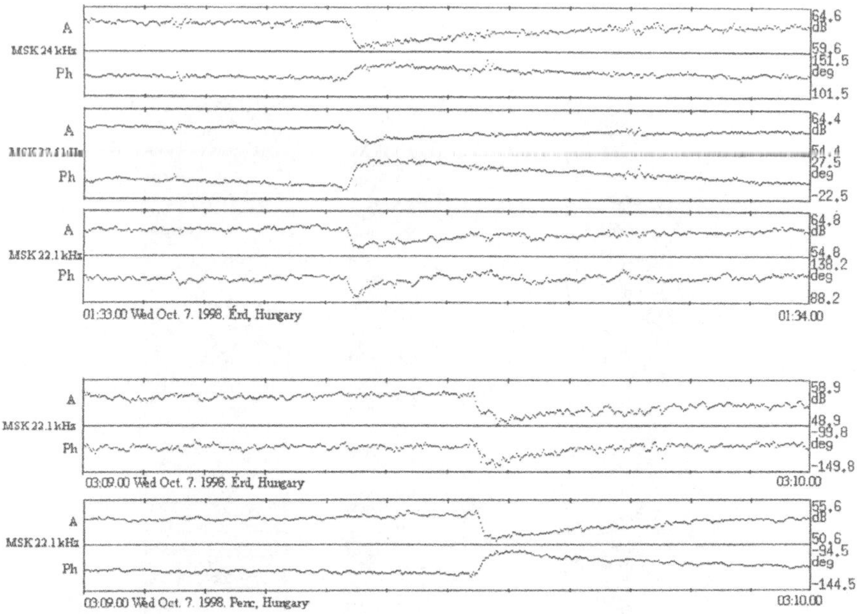

Figure 3.2. Trimpis recorded simultaneously at Érd and Penc, Hungary on 7 October 1998. (A – is the amplitude, Ph – is the phase)

3.2 Satellite measurements

The basic goals are the same in the case of satellite measurements as those of ground based ones. However, in space all the six field components can be measured, thus, beside the previously mentioned characteristics, the wave normal and the Poynting vector can also be calculated giving the possibility of analyzing all the aspects of wave propagation.

However, the in-situ measurements are limited by other constraints: opposite to the ground based measurements, here the size, the mass, the power consumption and the data transfer capacity of the measuring system are the basic limiting parameters. The satellite measurements in this field started more than 30 years ago; in the beginning, the analogue recordings and transmission were used, nowadays, digital recording and telemetry are common.

The additional phenomena that cannot or can hardly be detected on the surface are the different signals induced by physical processes in the plasma and this is a special benefit.

The magnetic field components are usually measured by light weight magnetic sensors (solenoids), usually shorter than 1 meter (a few dm) in length and smaller than 10 cm (a few cm) in diameter. The electric sensors consist of a pair of conductive spheres with a diameter of a few tens of centimetre (or around 10 cm) and several meters (or 1~2 m) in distance between them. During the history of satellite ELF-VLF measurements, the sensors also evolved. In early times, the magnetic sensors were able to cover only parts of ELF-VLF bands, e.g. 0-2.5 kHz on board GEOS-1, AUREOL-3 or only a few tens or hundred Hz wide band (DE-1, Cosmos1809, AKEBONO), in the better cases, usually more than one sensor was used. Recently, wide band (from cca. 1 Hz to 20~40 kHz) magnetic sensors are available for satellite measurements with enough sensitivity and signal-to-noise ratio (SNR) characteristics. The electric sensors were able to cover the ELF-VLF band from the beginning, thus the weakness of the electric sensors is not the bandwidth, but their sensitivity to polarization charges, that come either from the instrumentation of the whole space vehicle or from the ambient plasma.

The satellite can be three-axis stabilized or spin-stabilized. In the latter case, the characteristic time of the phenomena being investigated is (or close to) the same order of the spinning time, thus this effect has to be take into account. This rotation effect was incorporating e.g. during the analysis of the wave normal of signals measured on board DE-1 satellite.

Though it does not affect the measurements themselves, it should be mentioned here that in some satellites there were (e.g. ACTIVE, APEX) and will be (e.g. WARNING) – besides the passive probing of the ionosphere-magnetosphere system by ELF-VLF waves – active wave experiments too. In these experiments, the transmitted signals are received on the ground, on board the satellite and on board other satellites too.

Beside the similar events those detectable on the ground (Fig 3.3), unknown or not well known signals (Fig 3.4) can also be detected on board satellites.

Description of e.m. wave experiments can be found e.g. *S300 experimenters* (1979); *Kimura et al.*, (1980); *Berthelier et al.*, (1983); *Lichtenberger et al.*, (1991). One of these experiments was the successful operation of the first SAS (Signal Analyser and Sampler) instrument on board the ACTIVE (Intercosmos-24) satellite.

The SAS experiment was a joint effort of the Department of Geophysics, Eötvös University and the Department of Microwave Telecommunication, Technical University of Budapest. The technical preparation was carried out at the Technical University of Budapest in co-operation with IZMIRAN, Moscow. The scientific objective of the experiment was to study whistler/VLF propagation, VLF duct structure and the hyperfine structure of whistlers. In what follows, we briefly describe the SAS experiment.

The ACTIVE spacecraft was launched on 28 September 1989, from the (former) USSR into a moderately eccentric Earth orbit with initial perigee and apogee altitudes of approximately 500 km and 2500 km, respectively, an orbit inclination of 82.6°, and a period of 115.7 min.

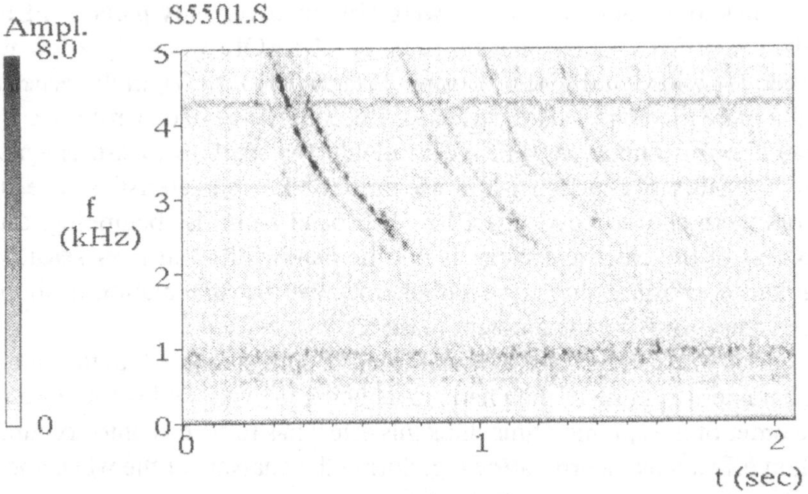

Figure 3.3 Whistler doublets recorded by SAS experiment on board the ACTIVE (Intercosmos-24) satellite on 14 December 1990, at 13:14:10 UT.

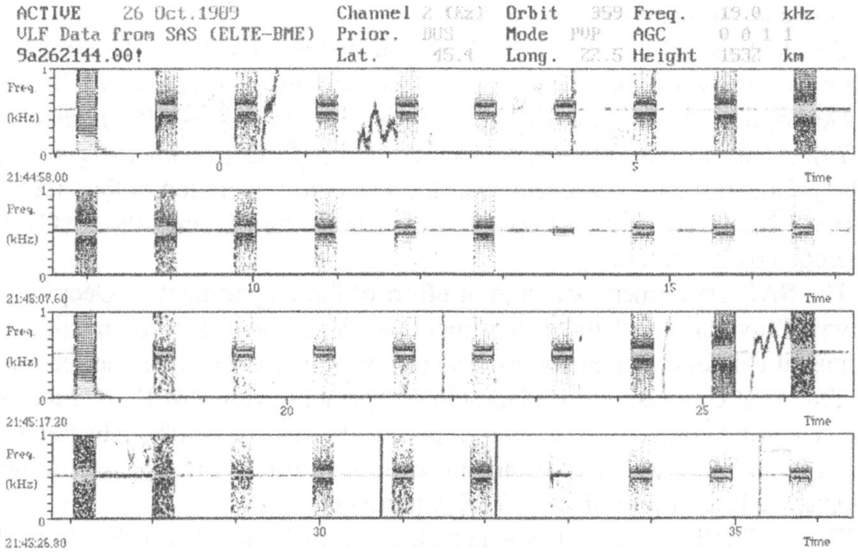

Figure 3.4. Strange signals recorded by SAS experiment on board the ACTIVE (Intercosmos-24) satellite on 26 October 1989, at 21:45 UT.

The SAS experiment was planned to measure the propagation characteristics (wave normal direction, polarization, amplitude) and the fine-structure of discrete VLF signals (mainly whistlers) by means of onboard five-component digital VLF waveform measurements and sophisticated data processing techniques. The SAS instrument consists of two blocks: the data processor block (DA) and a telemetry transmitter (P460). The DA block was originally designed to detect and process only useful signals (e.g., whistlers or other deterministic, discrete signals), but later, financial problems prevented us from realizing the onboard signal identification and solution function. The DA block receives three orthogonal magnetic and two orthogonal electric VLF field components from the onboard VLF-complex NVH-ONCH, built by IZMIRAN (Moscow). In the narrow-band mode, these input VLF signals are band limited with a bandwidth of approximately 900 Hz with a center frequency selectable in steps of 500 Hz between 500 Hz and 21.5 kHz. In the wideband mode, only a single (electric) component is received by DA in the band 20 Hz to 22 kHz. In narrow band mode, each of the received waveforms is sampled digitally with a rate of 2 kHz, and then 512-byte data frames are formed, which also include the necessary auxiliary information (time etc.). In the wideband mode, the sampling rate is (only) 10 kHz, which results in an increased noise level due to strong spectral overlapping, but otherwise the waveform of signals below 5 kHz is not distorted. The DA block also receives from and transmits to the NVK-ONCH various control data. It also contains a 64 Kbytes memory for the recording of waveform data over a period of ~6s anywhere along the orbit. The digital waveform data from DA are transmitted (usually in real-time) to ground receiving stations via the 460.4 MHz P640 telemetry transmitter, with an output power of 2.5 W per 50 ohms, phased shift (PSK) keying modulation and a data rate of 80 kbit/s non-zero returning.

SAS has three basic modes of operation: calibration, working with NVK-ONCH priority and SAS priority. In the second mode, SAS serves only for the sampling and transmission of waveforms in various experiments performed by the NVK-ONCH complex. In the case of SAS priority, which is the fundamental mode of operation of SAS, various constant and stepped frequency sampling and transmission modes can be set via the ACTIVE operation centre (USSR) or the command telemetry transmitter in Panská Ves, Czech Republic. Most frequently the 20 Hz to 900 Hz, 2.5 kHz to 3.5 kHz and 20 Hz to 5 kHz VLF bands were transmitted to the Earth. The data were received and immediately decoded mainly at the Technical University of Budapest. Altogether, about 180 Mbytes of information has been recorded on 150 satellite passes between 26 October 1989 and 31 January 1991. For 17 ACTIVE passes SAS real-time data were also received at NASA Goddard Space Flight Center's Wallops Flight Facility, Wallops Island. In addition to various onboard wave and particle measurements, the SAS measurements were accompanied by simultaneous ground-based VLF recordings at Tihany (Hungary), at Halley Station (Antarctica, by the British Antarctic Survey), in northern Finland (by the Geophysical Observatory, at Sodankylä) and in Grocka (former-Yugoslavia, by the Geomagnetic Institute). The joint analysis of ground-based and onboard VLF recordings will be interesting too. SAS operated properly till the end of 1992.

Automatic whistler detectors

One of the most severe bottleneck of the continuous ELF-VLF recording on the ground and on board satellites is that the pre-selection of data that contains interesting events requires continuous human attention. This simply means that scanning through one hour of registration requires (minimum) one hour listening of data. Because of this limitation, the common policy among ELF-VLF observations is to record periodically e.g. 2 minutes in every hour. The continuous or quasi continuous measurements are only made during measuring campaigns.

On board satellites this problem is connected to the limited size of board data storage capacity and the telemetry speed limitations too. The solution of this problem – at least in the case of *deterministic* signals – can be an automatic detector for the given phenomena, e.g. whistler detection, event detection. In the ELF-VLF range, the obvious candidates are whistlers for automatic detection. Though except some early, unsuccessful trials, operative automatic detectors were made only recently, simply because the speed of data processing of a complex signal reached the necessary level recently. The detection is not easy because of noise and – on the ground – particularly because of spherics.

One approach of detectors is using a neural network implemented by DSP. This method was applied to AUREOL-3 data (*Miniere et al.,* 1996) and will be used on board the DEMETER satellite (*Elie et al.,* 1999).

Another approach is using frequency discrimination filter. This method is used at Tihany observatory (Hungary) from the end of 2000 for ground based measurements and in SAS2 equipment on board COMPASS and PRECURSOR satellites.

This automatic equipment can detect whistlers (or deterministic events), record the time of the event and store only that part of data that contain whistlers (or events). This way, not only the tiresome human attention is eliminated, but the required data storage and telemetry speed is considerably reduced. Besides these, the automatic whistler/event detection opens the way for *non-stop* recording, i.e. for real e.m. monitoring.

3.3 Goals of future ELF-VLF investigations

In the atmosphere of the planets and in the interplanetary space significant ELF-VLF electromagnetic activity exists. The sources of this activity are different, the shape of these signals can be arbitrary. As was seen earlier, we have models and some interpretations for a few types of these signals, first of all for the whistler-type signals. However, we have a lot of satellite and on the ground recorded signals with strange shapes (see e.g. Fig. 3.4 and 3.5 or Fig. 3.6) and without any acceptable and confirmed explanations (e.g. *Tarcsai et al.*, 1992).

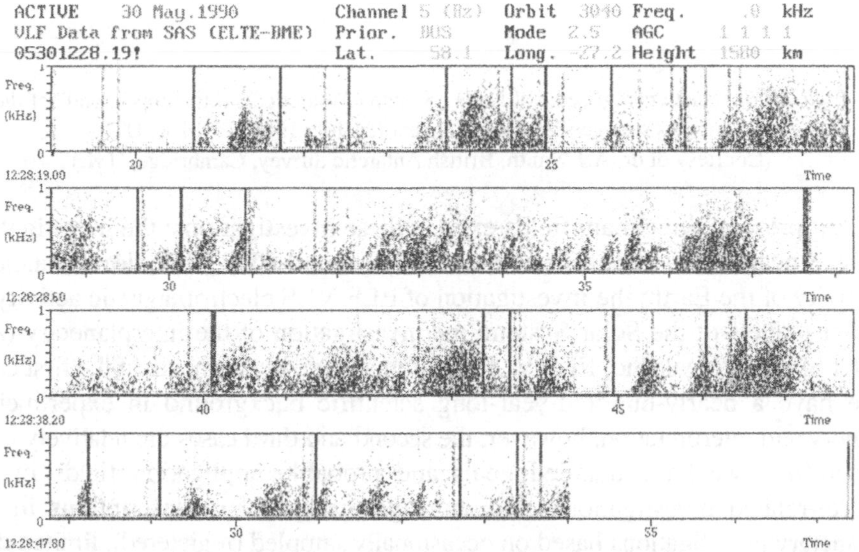

Figure 3.5. Strange shaped signals with unknown origin registered by the SAS instrument on board ACTIVE (Intercosmos-24) satellite (30 May 1990, 12:28 UT).

On the other hand in the last years, it was clarified that the ELF-VLF signals give an excellent tool for investigation and monitoring of the magnetoionized part of the planetary (including Earth's) atmosphere, for investigation of the atmospheric activity of planets (including Earth) (e.g. *Jiricek and Tarcsai,* 1980; *Thomson and Dowden,* 1977a,b) and we have remarkable indications of a probable relation between the ELF-VLF electromagnetic activity and the seismic events of a given planet (e.g. *Hayakawa and Fujinawa,* 1994; *Hayakawa,* 1999). Therefore the ELF-VLF electromagnetic investigations have an important role in future space research and applications. The most probable and most important fields and goals of space activity will be presented in the following.

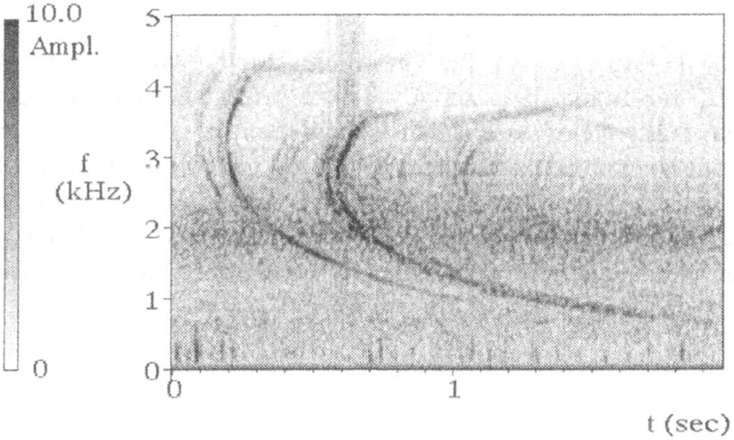

Figure 3.6. Whistlers registered at Halley Station (Antarctica), at the upper "end" of the whistlers strange signal-parts appear (21 July 1984, 17:14:14 UT). (Courtesy of dr. A.J. Smith, British Antarctic Survey, Cambridge, U.K.)

At present, three main fields exist in these investigations: the investigation of the near-Earth, atmospheric and at the ground ELF-VLF electromagnetic activity of the Earth; the investigation of ELF-VLF electromagnetic activity of other planets of the Solar System; the investigation of the interplanetary (and later of the intergalactic) ELF-VLF electromagnetic phenomena. In the first case, we have a nearly-hundred-year-long scientific background in experiments, theory and interpretation, however, the second and third cases are relatively new. Therefore, we have detailed goals and probable application fields in the Earth-related investigations, and more general problem descriptions in the planetary investigations based on occasionally sampled (registered), limited data set.

Besides these, fields special ELF-VLF measurements open the way to control, confirm or clarify problematic questions in electromagnetic wave propagation theory using space based experiments.

3.3.1 Experimental investigation of theoretical questions of electromagnetic wave propagation

In the theory of electromagnetic wave propagation, the propagation of the energy, signal maxima, amplitude-changes and perturbations is clear without any contradictions in vacuum and in a few, very simple cases, as for example a medium with a simple polarisation interaction between the signal and the medium. However, in more complicated cases (propagation in moving media, in inhomogeneous, magnetoionic media etc.) some contradictions appeared in the theoretical results. E.g. the widely known contradiction between the transformation of signal velocity and the Einstenian addition of velocities in simple moving dielectrics or in more complex media. Different experiments are known for the resolution of these problems. These problems are in a direct connection to the definitions of the propagation factor of the energy $(\bar{k}_g)$, to the definitions of the energy or group or wave-front velocities ($\bar{v}_g$ etc.), to the definitions of the direction of these vectors and of the relation of these vectors to the Poynting-vector $(\bar{S})$ and the field vectors $(\bar{E}, \bar{H}, \bar{D}, \bar{B})$ and the normal unit vector of the wave-front surfaces $(\bar{n}_0)$. See the earlier discussion of these problems e.g. in *Laue* (1955), and the recent results in this field in *Ferencz* (1977b, 1978b, 1987, 1996). Our well based opinion is that energy propagation vector $(\bar{k}_g)$ exists, but in the case of a wave-front or other wave-like (i.e. not point-like or particle-like) formation energy velocity vector $(\bar{v}_g)$ does not exist, only it is possible to define a v_g number using the absolute value of $\bar{k}_g$. On the other hand, the direction of the energy propagation factor $(\bar{k}_g)$ and the real propagation direction of a wave-front is parallel to the normal unit vector $(\bar{n}_0)$ independently of the other vector directions which can be calculated using the field vectors $(\bar{E}, \bar{H}, \bar{D}, \bar{B})$.

Space research gives an excellent opportunity to control the different theoretical interpretations and to decide between these using direct experimental data. The International Space Station (ISS) is the first space vehicle which has so large measures that the time delay of a propagating signal in ELF-VLF bands between the different points of the structure is well detectable with high accuracy. In the Earth's atmosphere the propagation velocities of ELF-VLF signals at the orbit of ISS produce time delays with measurable values on the distances between the farthest points of ISS. Thus, an acceptable measuring configuration is the following:

Let us use four measuring (signal detecting) instruments at the far end points of this big space-structure (see e.g. the Fig. 3.7). These units detect the accurate transition times of a given signal at the units. These data determine in an elementary manner, the real position of the wave-front plane in space and therefore the normal unit vector $(\bar{n}_0)$ of the wave-front, the propagation vector $(\bar{k}_g)$ of the signal and the value of the equivalent propagation velocity $(\bar{v}_g)$. On the other hand, it is possible to measure the six field-components of the electromagnetic signal – three electric and three magnetic components – with one or with all four measuring units. (In general, it is enough to measure the six components with one unit only.) These data define the Poynting-vector $(\bar{S})$ and from this data group it is possible to calculate signal-plane using theoretical models. The comparison of the two data groups gives the opportunity of control and confirmation of the present existing theoretical ideas. Another benefit of this measuring system is, that the produced data set can calibrate the data of the signal propagation directions, which are extremely important in possible monitoring given in details in points 3.3.2 and 3.3.3.

It is evident that this investigation can also be made by using three or four satellites in a simultaneous measuring mode, flying in the same space-segment above the Earth surface.

3.3.2 Investigation of possible relationships between seismic activity (precursors and earthquakes) and ELF-VLF phenomena

At the moment of closing this manuscript, the relationships between the seismic activity (of the Earth and later of other planets with solid surface) and the ELF-VLF electromagnetic activity of that planet detectable on board satellite only are probable and possible. However, the importance of this possibility is clear.

The preliminary investigations of earlier satellite data and some hypothetical theoretical modelling give a few possible phenomena inside this field of investigations (see e.g. *Hayakawa and Fujinawa,* 1994; *Hayakawa,* 1999). The common part of these possible phenomena is that the hoped ELF-VLF electromagnetic indicator signals have an arbitrary shape and we have no preliminary information about the signal form. If the following research find relations between ELF-VLF signals and seismic activity, the next step of investigations will be the theoretical modelling of the source signal generation and computing the propagation from the source to the satellite using the methods presented in Chapters 1 and 2.

In this phase of research, we see the following main groups of possible events, based on earlier indications (see e.g. in *Dziewonsky et al.*, 1971: *Hayakawa and Fujinawa,* 1994; *Hayakawa,* 1999; *Parrot,* 1994):

– increasing the general, noise-like ELF-VLF and HF-UHF electromagnetic activity and occurrence of electromagnetic (noise) bursts;

– occurrence of special deterministic e.m. signals (with anomalous shape) as precursors or indicators of (stronger) seismic activity;

– changing in the characteristics of known atmospheric ELF-VLF phenomena, e.g. whistlers appearing with extremely large dispersion;

– local distortion of the ionosphere, which changes the parameters of the surface – ionosphere wave guide and therefore the characteristics of the signals of ELF-VLF transmitters, e.g. the phase-curves of these signals will change – which is a new direction in trimpi research;

– appearance of transient (heath) anomalies in the thermal infrared (and in the optical and microwave) remote sensing data (pictures) of the Earth's surface, which is a new direction in satellite remote sensing. However, this is out of the scope of this presentation.

Figure 3.7. Positions of signal measuring units (the four SAS3 devices) on the International Space Station

At the moment (at the end of 2000) some satellite projects exist to investigate the ELF-VLF activity ⇔ seismic events relations. The most important are the COMPASS and PRECURSOR satellites, the DEMETER, the WARNING, the CESAR and the Global Electrodynamics Explorer projects.

3.3.3 The investigation of whistlers and the general ELF-VLF electromagnetic activity around the Earth

There are open questions in whistler research. In the near future the investigation and monitoring of the plasmasphere by whistlers, the investigation of the occurrence of strange-shape signals and the monitoring of man-made effects will be important. In the following, we will present the most important items of this activity.

– Study of whistler's *L*-values discrepancy: It has been shown earlier (e.g. *Thomson and Dowden,* 1977a,b; *Jiricek and Tarcsai,* 1980) that the L-values of propagation of the "ducted" whistlers observed aboard satellites are rather independent of the L-value of the satellite position itself. Similarly constant dispersion whistlers were observed over wide latitudinal ranges aboard low altitude satellites. It is still not clear which mechanism is responsible for the L-value discrepancy (see in part 2.5). One possible explanation is that the signal exits from the duct at an altitude well above the satellite. It is also possible that the downgoing whistler mode signals reflect from the ionosphere at a level well below the satellite. Other theories suppose that these whistlers are the so-called pro-longitudinal ones, which do not follow field-aligned paths. Calculating accurate angles (directions) of propagation of observed whistlers can prove or disprove these theories.

– Study of fine structure of whistlers: By means of accurate wave analysis (e.g. *Tarcsai,* 1975; *Gendrin and Villedary,* 1979; *Hamar and Tarcsai,* 1982; *Hamar et al.,* 1990; *Lichtenberger et al.,* 1991), it has been proved that whistlers which appeared as well defined, discrete, single traces on a conventional spectrogram are regularly composed of several components covering various parts of the spectrum (see more in Chapter 4). This detailed structure can be explained by the multiple stroke of the causative spherics, the effects during the ducted or non-ducted propagation in the magnetosphere or mode splitting. The detailed analysis of whistlers using e.g. the data of an onboard automatic whistler detector and comparing these to results of an extremely high time resolution wave pattern detector data can answer these questions.

– In whistler research, it is a common opinion (conclusion) that a lot of whistlers or the dominant part of whistlers recorded at the ground stations and on board satellites are ducted signals, i.e. these signals propagate along a magnetic field-line during the whole propagation or during the section of the path which causes the dominant part of signal dispersion. This common opinion is based on some model deductions based originally on monochromatic-quasimonochromatic theoretical approximations. (This is the main reason why in Chapters 1 and 2 we investigated a lot of cases of longitudinal propagation.) The deducted effective "diameters" of the ducts are between a few tens km and a few hundred km. However, if these ducts exist, an ELF-VLF signal can propagate in these ducts only in wave-guide modes, i.e. the duct is a waveguide, such as known in microwave techniques and in fiber-optics too (see e.g. *Simonyi,* 1963). In these propagation modes we have a limiting factor of the propagation, the limiting wavelength. The value of the limiting wavelength depends on the size of the waveguide (duct) and on the actual guided mode of the signal. The signal cannot propagate above this limiting wavelength, i.e. below this limiting frequency. On board space vehicles the general noise level in the ELF bands is small and therefore it is possible to control that the lower asymptote of a given whistler in the dynamic spectrum is the zero frequency or a higher value. If the signal propagated in a duct, the value of the asymptote must be definitely higher than zero.

– Earlier wave analysis of the SAS experiment on board ACTIVE (Intercosmos-24) satellite shows special patterns of some signals in the dynamic spectrum (e.g. *Tarcsai et al.,* 1992, or Figs. 3.4 and 3.5). The fine analysis of the observed signals by satellite measurements gives a possibility of the study of such signals. The theory of propagating electromagnetic waves in magnetized plasma can be developed and further improvements of the model calculations could be made to give explanations of propagation characteristics of these very strange signals.

– Study of noise-like (diffuse) whistlers and other man-made effects: Whistlers recorded earlier (e.g. in the beginning of 70's) in some observatories (e.g. Tihany, Hungary; Panská Ves, Czech Republic; Sodankylä, Finland; Poitiers, France) show, in most cases, well defined, discrete single traces on a conventional spectrogram (see e.g. in Fig. 3.1b). Even these signals may consist of several, closely spaced traces, they significantly differ from those ones observed at the same observatories after the mid-80's. The signals that could be recorded recently in these locations, in most cases, show like a 10-50 ms broad, dispersed noise. To determine a single, dispersed trace with at least several hundred Hz bandwidth even with the most sophisticated procedure is rather difficult, and sometimes impossible. However, well-defined traces can be found in whistlers recorded on the Halley Station, Antarctica or Banaras, India or on board satellites. Simultaneous onboard and ground observations can give explanation of these phenomena and determine the location of the territories where noise-like whistlers (or whistler-like noises) can be observed. The possible connection of the occurrence of noise-like whistlers and human industrial activity can be examined. Besides these investigations, it is also possible to measure the occurrence of non-common shaped signals and to monitor the general level, structure and spectrum of the ELF-VLF noise. The third part of this research is the simultaneous monitoring of the plasmaspheric parameters (e.g. by whistlers) and the analysis of the parameter variations. This group of investigations can give some answer on the effects of man-made variations in the higher atmosphere of Earth and on the fact and level of ELF-VLF electromagnetic pollution.

3.3.4 Investigation of ELF-VLF activity of other planets

The main goals of these investigations and the possible mechanisms of the ELF-VLF activity of other planets are similar to the case of Earth, in general. An important difference is the case in which the planet of interest has no (remarkable) magnetic field, see e.g. the planet Mars or the Moon. In this last case the planet has no plasmasphere like the Earth or the big planets of the Solar System. However, ELF-VLF activity must be produced by the planets and the bigger moons. In the past (and present) some ELF-VLF measurements were made around and at the vicinity of planets (e. g. *Smith et al.*, 1980; *Kaiser et al.*, 1980; *Gurnett et al.*, 1990; *Menietti et al.*, 1991; *Sonwalkar et al.*, 1991).

In the case of a planet that has magnetosphere, it is possible to monitor the whistler propagation, as is common in the Earth related research and applications. This gives a good opportunity to monitor the plasmaspheric parameters and conditions of the given planet, e.g. in the case of the big planets of the Solar System.

Independently of the existence of plasmasphere, every planet has a significant atmospheric activity (storms etc. – see e.g. Venus, Mars, big planets), remarkable seasonal changing in some cases (e.g. Mars) and ionosphere. A detailed investigation of the ELF-VLF environment of those planets and the analysis of the detected signals will produce important new information about the items mentioned above.

From another point of view, the ELF-VLF monitoring and mapping of the other planets will give new information not only about the atmospherical-plasmaspherical environment, but this mapping and monitoring could produce important information about the seismic activity of the planets too.

The possible signal detection goals used in these ELF-VLF monitoring could be the following:

– Triggered event detection: in this case the measuring instrument records an incoming ELF-VLF event, if the computed spectrum of the input signal differs significantly from the flat noise spectrum. This is important, because we have no good overview on the nature of e.m. activity of other planets.

– By using a special whistler-detector (or spherics-detector) unit, it is possible to monitor and map the whistler (and lightning) activity of a planet.

– The variations of the general ELF-VLF longer time averages, i.e. the variations of the average noise level could be informative too.

3.3.5 Investigation of interplanetary space

The Sun and the planets are real sources of e.m. signals and a lot of these are inside the ELF-VLF bands. The signals propagate through the interplanetary plasma and the detailed investigation of the shape of these signals could produce information about the propagation, i.e. about the interplanetary plasma itself, and about the source, the origin of these signals. This means a triggered event detection as was mentioned in point 3.3.4, and the registration of the average noise level variations is also important. In the farther future similar measurements will be important also in the interstellar space.

We can say that in the investigations presented in points 3.3.4 and 3.3.5 the sources of the essentially new information are

– the long time monitoring (and mapping) of the general ELF-VLF electromagnetic activity,

– the detection, identification, analysis and modelling of strange shaped signals and

– the investigation of signal sources and traversed medium by analysis of known signals, e.g. by whistlers.

Chapter 4

METHODS OF SIGNAL ANALYSIS

The method using whistlers (impulse-generated signals) is one of the small number of techniques which yield reliable data on morphology and dynamics of the medium traversed by the signal, e.g. of plasmasphere, on magnetospheric electric fields and of ionosphere-magnetosphere coupling fluxes of the Earth and other planets. On the other hand studying the signal waveform, their amplitude-frequency behaviour the better understanding of the wave propagating in the upper atmospheric plasma can be achieved.

For the analysis of measured signals and studying the model-calculated signals the accurate determination of the frequency-time-amplitude characteristics of these signals is obvious. The different techniques we outline here concentrate on the analysis of electron-whistlers but these techniques can be adapted to any discrete signals with arbitrary shape, i.e. with arbitrary frequency and amplitude variation.

4.1 General character of the signals

The usual methods of spectrum analysis are not suitable for the high accuracy analysis of time-varying general shaped signals, which are not stationary. However, the further study of such signals frequently necessitates the accurate determination of the dynamic spectrum or the variation of frequency in time. To overcome this difficulty and also to increase the speed of data processing, different numerical methods have been developed in addition to the Sonograph or Rayspan type analogue technique and stationary spectrum analysis.

The simplest way to build up a "digital sonogram" is to apply a moving time – or frequency – window to the properly sampled signal and then to compute the amplitude of the dynamic spectrum and plot it as a function of two discrete variables f and t. A moving time-window has been used in conjunction with Fast Fourier Transform (FFT) procedures for the analysis of VLF signals first by *Coroniti et al.* (1971), who used a digital spectrum-analyzer, *Cöster* (1974) and *Stiles and Helliwell,* (1975). However, it is well known the relation of indeterminacy, that if we accurately specify the frequency (narrow frequency window) we can determine the time of existence only less accurate, and dividing more finely the time axis (narrow time window) the momentary signal-frequency will be less accurate.

Another way to find more accurately the *f-t* behavior of a given signals is to use filters of optimum bandwidth, which requires a priori knowledge of the approximate *f-t* function. *Bashilov et al.* (1977), *Mihajlova and Kapustina* (1978) and the FTAN algorithm developed by *Levshin et al.* (1972) used optimum bandwidth Gaussian filters. They analyzed electron whistlers and proton whistlers (below 1 kHz) and achieved a much higher accuracy in determining the frequency-time and amplitude-time relations than that of the classical methods of dynamic spectrum analysis.

A certain class of methods was based on the classic paper by *Gabor* (1946), who gave an interesting representation of signals with time varying frequency. He proposed to represent the signal as the sum of "elementary signals", which are limited both in duration and frequency content. A generalization of *Gabor's* work has been made by *Lerner* (1961) and *Daniell* (1969, 1970). They proposed algorithms to compute the complex amplitudes of elementary signals for the construction of the *f-t* plane. Some examples obtained with this method were presented by *Daniell and Papanicolau* (1975). *Helstrom* (1968) and others represented the signals by integrals over a continuous frequency-time plane.

Griffiths (1975) derived the instantaneous frequency content of a signal from a spectral computation based upon the use of an adaptive linear prediction filter. The main advantage of the procedure is that it is well suited to real-time applications.

In the procedures outlined above the use of the digital technique enables a frequency-time analysis of whistlers and other time-varying signals which is faster and generally more accurate than the conventional analogue methods of spectrum analysis. The accuracy can be further increased by the application of correlation detection or matched filtering. The method developed and widely used in the radar technique (see e.g. *Cook,* 1960; *DiFranco and Rubin,* 1968) was applied for whistler first by *Béghin and Siredey* (1964).

The matched filter technique was applied to whistlers recorded digitally at Halley, Antarctica (*Hamar et al.,* 1990), where the obtained *f-t* pairs with the corresponding magnitudes gave a high resolution dynamic spectrum, which revealed the fine structure of whistlers. Whistlers, which appeared as well defined, discrete, single traces on a conventional spectrogram, turned out to be composed of several components covering various parts of the spectrum. The matched filter method of analysis could provide new insights into the details of whistler and pulse propagation. It also yielded the amplitude variation vs. time along a signal (whistler) trace.

4.1.1 The matched filter

Consider the problem of specifying the impulse response of a linear time-invariant filter i.e. the optimum filter, $h_{opt}(t)$, which maximises the output signal-to-noise ratio (SNR) at time $t = t_m$. The input waveform is $u(t) = s_i(t) + n_i(t)$, where $s_i(t)$ represents the signal component, $n_i(t)$ the noise component, while the output waveform is $y_0(t) = s_0(t) + n_0(t)$. The output SNR is defined as

$$\chi(t) = \frac{s_0^2(t)}{\langle n_0^2(t) \rangle}, \tag{4.1}$$

where $\langle\ \rangle$ means the average values. If the optimum filter has a finite length (or duration) *T*, the noise is additive and Gaussian, with a two-sided constant power density $N_0/2$, an equation can be derived for $h_{opt}(t)$ (see e.g. *Zadeh and Ragazzini,* 1952; *DiFranco and Rubin,* 1968)

$$h_{opt}(t) = \frac{2k}{N_0} s_i(t_m - t), \quad 0 \le t \le T, \tag{4.2}$$

where *k* is constant. This equation shows that for a known signal $s_i(t)$ of finite duration *T*, the impulse response of the optimum linear filter is proportional to the input signal with reversed time axis. This filter is called a "matched filter".

Let the time autocorrelation function $\Phi_{ss}(\tau)$ of $s_i(t)$ be defined by

$$\Phi_{ss}(\tau) = \int_{-\infty}^{\infty} s_i(v) s_i(v - \tau) \mathrm{d}v.$$

Then the output signal can be rewritten as

$$s_0(t) = \int_{-\infty}^{\infty} s_i(v_i) h_{opt}(v - t) \mathrm{d}v = \frac{2k}{N_0} \Phi_{ss}(t - t_m). \tag{4.3}$$

Thus the output signal of a matched filter is proportional to the autocorrelation function of the input signal. The average output noise power of a matched filter will be

$$\langle n_0^2(t_m)\rangle = \frac{2k^2}{N_0}W, \tag{4.4}$$

where W is the input signal energy, i.e.

$$W = \int_{-\infty}^{\infty}[s_i(t)]^2\,\mathrm{d}t = \Phi_{ss}(0).$$

The peak output SNR follows from equations (4.1), (4.3) and (4.4), which yield at $t = t_m$:

$$\chi(t_m) = \frac{2W}{N_0}.$$

Although the waveform of whistlers propagating through the plasmasphere are usually rather complicated, it is useful to consider the application of a matched filter to a linear frequency-modulated pulse. Later we can prove that the conclusions derived from matched filtering of a sweep can be adapted to whistlers.

4.1.1.1 Matched filter approximation for a linear frequency modulated pulse with rectangular envelope

Let the input signal $s_i(t)$ be a pulse of amplitude A, duration Δt, whose frequency changes linearly over Δt as

$$s_i(t) = A\cos\left(\omega_m t + \frac{1}{2}bt^2\right),\quad t_m - \frac{\Delta t}{2} \le t \le t_m + \frac{\Delta t}{2}. \tag{4.5}$$

The instantaneous frequency at the center $(t = t_m)$ of the pulse is $f_m = \omega_m / 2\pi$ and the magnitude of the linear frequency sweep is $\Delta f = b\Delta t / 2\pi$. In the following section we shall see that even the full-wave solution describes the whistler waveform precisely, for studying the matched filtering technique the whistlers can be approximated by the above waveform, therefore it is expedient too recall here a number of relevant, useful formulae to be referred to later in this chapter.

The output signal of the matched filter is obtained by substituting (4.5) into (4.3). If the length of this (frequency modulated, i.e. FM) pulse – see relation (4.5) – is much longer than the period of the carrier wave $(\Delta t \gg 1/f_m)$, which is valid in most applications, and for $|t'| \ll \Delta t$, that is vicinity of $(t = t_m)$, the output signal reduces to (*DiFranco and Rubin,* 1968)

$$s_0(t) = \frac{kA^2\Delta t}{N_0}\text{sinc}(\Delta f t')\cos(\omega_m t'), \qquad \text{where} \qquad t' = t - t_m \tag{4.6}$$

and $\text{sinc}(x) = \sin \pi x / \pi x$ by definition. Thus the output is a cosine signal of frequency f_m, modulated in amplitude by a slowly varying sinc function, the first zeros of which are at $|t'| = 1/\Delta f$. The half power width of the central peak of the envelope is also of the order of $1/\Delta f$. The peak value of $s_0(t)$ occurs at $t = t_m$,

$$s_0(t_m) = \frac{kA^2\Delta t}{N_0}. \tag{4.7}$$

Hence, the epoch of the main peak coincides with the centre of the input FM pulse, where its instantaneous frequency is equal to f_m. If the output is strongly compressed, the position of its centre is well defined and the time of arrival of the input pulse can be precisely obtained. This principle of pulse compression was introduced first by *Klauder et al.* (1960) in the radar technology and has been widely used since that time with various pulse compression waveforms. The peak output SNR is given by

$$\chi(t_m) = \frac{2W}{N_0} = \frac{A^2\Delta t}{N_0}. \tag{4.8}$$

It is important to note that χ depends on the power density of the input noise and is not affected by its total power. Therefore it is unnecessary to increase the input SNR by using a band-pass filter around f_m prior to matched filtering, because it does not enhance the output SNR.

It is interesting to compare $\chi(t_m)$ to the input SNR, $\chi_i(t_m)$, which is defined according to equation (4.1) as

$$\chi_i(t_m) = \frac{s_i^2(t_m)}{\langle n_i^2(t_m)\rangle}. \tag{4.9}$$

Because of the mean of the input noise values equals to zero, the σ^2 variance of the noise will be $\sigma^2 = \langle n_i^2(t_m)\rangle$. On the other hand, if there is a white Gaussian noise with standard deviation σ, then $\sigma^2 = N_0 B$, where B is the noise bandwidth. Hence $n_i^2(t_m) = N_0 B$. The largest possible input SNR occurs at the peaks of the oscillating input signal, when $s_i(t) = \pm A$ and

$$\chi_i(t_m) \leq \frac{A^2}{N_0 B}. \tag{4.10}$$

From (4.8) and (4.10) it can be seen that the relation of output SNR to the input SNR is given by

$$\chi(t_m) = B\Delta t\, \chi_i(t_m). \tag{4.11}$$

From (4.11) it is clear that the output SNR $\chi(t_m)$ is a linear function of the length (Δt) of the filter.

Because of the random noise component of the output waveform $y_0(t)$, the actual epoch $\hat{t}_m$ of the main peak may deviate from t_m. According to *Woodward and Davies* (1950), the standard deviation of $\hat{t}_m$ around t_m is

$$\sigma_{t_m} = \frac{1}{\beta\left(2\frac{W}{N_0}\right)^{1/2}} = \frac{1}{\beta\sqrt{\chi(t_m)}}, \qquad \beta = \left[\left\langle(\omega-\omega_m)^2\right\rangle\right]^{1/2} \tag{4.12}$$

is the effective bandwidth of the filter, which is related to the linear frequency sweep Δf as $\beta = 1.8 \cdot \Delta f$ (*Rihaczek,* 1969). Because $\sigma^2 = N_0 B$ and applying (4.8), (4.12) can be rewritten as

$$\sigma_{t_m} = \frac{N_0^{\ 1/2}}{1.8 A\Delta f \Delta t^{1/2}} = \frac{\sigma}{1.8 A\Delta f\,(B\Delta t)^{1/2}}, \tag{4.13}$$

which shows clearly the effect of different input parameters on the uncertainty in $\hat{t}_m$.

4.1.2 Application of matched filtering to whistlers

According to the expression (4.2), the matched filtering requires the knowledge of the form of the signal to be filtered. Generally, a signal with frequency varying in time can be described in most simple form by

$$s(t) = \cos[2\pi g(t)], \tag{4.14}$$

where the instantaneous frequency is by definition

$$f(t) = \frac{\partial g(t)}{\partial t}. \qquad \textbf{(4.15)}$$

If the $s(t)$ signal is a whistler, we can determine by a simple technique (e.g. by scaling from the dynamic spectrum) a series of f-t pairs. However, to construct a matched filter the f-t pattern of the signal should be known more accurately. To increase the accuracy of the f-t data series, a method (i.e. the FIT method) has been developed by *Tarcsai* (1975) based on the least squares estimation of the parameters in *Bernard's* (1973) approximate function for the dispersion (the f-t pairs) of (measured on computed) whistlers.

4.1.2.1 The FIT method

For the routine application of a method which describes the approximate f-t pattern of a whistler, *Bernard's* (1973) approximate function has been used

$$D(f) \equiv t(f)\sqrt{f} = D_0 \frac{f_{HE} - Af}{f_{HE} - f}, \qquad \textbf{(4.16)}$$

where D_0 is the first, constant term in the Taylor-series of $D(f)$ derived from travel-time integrals (*Storey*, 1957), i.e. the "zero-frequency dispersion", f_{HE} is the equatorial electron gyrofrequency, f is the momentary wave-frequency, $t(f)$ is the "travel time" (time delay) at frequency f, and

$$A = \frac{3\Lambda_n - 1}{\Lambda_n(1+\Lambda_n)}, \quad \Lambda_n = \frac{f_n}{f_{HE}}, \qquad \textbf{(4.17)}$$

where f_n is the nose frequency. The nose frequency is the frequency in dynamic spectrum of the (measured or computed) signal having the smallest time delay – see e.g. the FFT pattern of measured signals in Fig. 1.5. From equations (4.16) and (4.17) the travel time at the nose frequency is

$$t_n = \frac{D_0}{\sqrt{f_n}} \frac{2}{1+\Lambda_n}. \qquad \textbf{(4.18)}$$

If the excitation signal, i.e. the causative spheric is unknown, and the travel times at different frequencies of the whistler trace are measured with respect to an arbitrary time origin, it is necessary to introduce a new (unknown) parameter T which gives the difference in time between the chosen time-origin and the actual excitation. Using T and equation (4.16), the measured travel time t^* can be written as

$$t^*(f) = t(f) - T = \frac{D_0}{\sqrt{f}} \frac{f_{HE} - Af}{f_{HE} - f} - T. \tag{4.19}$$

In this equation there are four unknown parameters: D_0, f_{HE}, T and A (or f_n). In the following we have two possible way deriving the signal parameters. Either A is considered to be actually an independent parameter, or adopting a model for electron density distribution we can derive a relation between f_n and f_{HE}, thus the number of unknowns can be reduced to three (as f_n and A depend on f_{HE}). In this method the latter possibility has been chosen. Thus, those values of D_0, f_{HE} and T are searched for, which fit best to the measurements in a least squares sense, i.e. which minimize the sum of the weighted squares of residuals

$$M = \sum_{k=1}^{n} w_k \left[t^*_{mk} - t^*_{ck}(D_0, f_{HE}, T) \right]^2, \tag{4.20}$$

where the subscripts m and c refer to the measured and computed t^* values, respectively, w_k are the weights given to the individual measurements, and the summation is to be taken over the points of the whistler trace scaled at frequencies f_k. For the sake of simplicity, in equation (4.20) the measurement errors were supposed to be uncorrelated.

As it is known from the theory of least squares estimation, in nonlinear case the determination of the parameters which correspond to the condition $M \triangleq$ *minimum* can be accomplished through linearisation by expanding the residuals into Taylor series and then using an iteration procedure (e.g. in *Hudson,* 1970). Exactly this has been done in this case also. Let us introduce the vector $\overline{X}$ whose components are

$$x_1 = D_0, \qquad x_2 = f_{HE}, \qquad x_3 = T. \tag{4.21}$$

Then assuming that the values of D_0, f_{HE}, and T are known at iteration $(i-1)$, the improved solutions at the i-th step can be obtained as

$$\overline{X}_i = \overline{X}_{i-1} + \Delta\overline{X}_i, \tag{4.22}$$

where the vector of differential corrections $\overline{\Delta X}$ is given in each step

$$\overline{\Delta X} = \overline{\overline{A}}^T \overline{\overline{W}}\,\overline{Y} \cdot \left(\overline{\overline{A}}^T \overline{\overline{W}}\,\overline{\overline{A}} \right)^{-1}. \tag{4.23}$$

In equation (4.23) the upper index T indicates matrix transposition, $\overline{\overline{A}}$ is the $n \times 3$ matrix of the partial derivatives of t^* with respect to the unknown parameters, whose elements can be computed using equation (4.19)

$$\sigma_{kj} = \left(\frac{\partial t^*}{\partial x_j} \right)_{f=f_k} , \tag{4.24}$$

where $\overline{\overline{W}}$ is the $n \times n$ square matrix measurement weights, which is diagonal for uncorrelated measurement errors, and $\overline{Y}$ is the column vector of residuals with elements

$$y_k = t^*_{mk} - t^*_{ck} . \tag{4.25}$$

In evaluating the matrix formula in equation (4.25), those values of D_0, f_{HE} and T are to be used, which has been obtained in the previous step of iteration and no weighting has been used in general ($w_{kk} = 1$, $w_{kj} = 0$ for $k \neq j$).

The value of f_n has also been improved in each step, in the following manner. According to *Park* (1972), for a typical diffusive equilibrium model (DE-1) of electron density distribution in the plasmasphere, Λ_n can be calculated from f_n as

$$\Lambda_n = \left(3.5475 - 0.47351F + 0.065879F^2\right)^{-1} \tag{4.26}$$

where $F = \log_{10}(f_n)$. Remembering that $\Lambda_n = f_n / f_{HE}$ we can write

$$f_{n,i+1} = \Lambda_{ni} f_{HE,i+1} \tag{4.27}$$

where Λ_{ni} is computed by equation (4.26), using f_{ni} obtained in the preceding step of iteration. Thus, for a converging f_{HE} the nose frequency also converges. Knowing these and discussing (4.17), it is clear that in the course of the iteration the value of A is also changed successively, but in general rather weakly, and it converges very fast.

The iteration procedure outlined above can be stopped if the magnitude of corrections decreases under a certain fixed level or the sum of the weighted squares of residuals stabilizes. After the criterion of convergence has been fulfilled, i.e. the iteration has been finished, t_n can be computed from equation (4.18).

Our experiences have shown that the described procedure generally converged quite well. However, to ensure the procedure against divergence it is advisable to introduce a multiplicator m into equation (4.22) (*Hudson*, 1970)

$$\overline{X}_i = \overline{X}_{i-1} + m\overline{\Delta X}_i, \qquad \textbf{(4.28)}$$

where $0 < m \leq 1$ and the actual value of m is properly varied in the course of the iteration. Generally $m \leq 0.1 - 0.3$ at the first two steps and in the following steps it is increased to unity.

4.1.2.2 Construction of the matched filter

By the relation (4.18) and the obtained D_0, f_{HE} and T outlined above we can calculate numerically the value of the $f(t)$ function – i.e. the instantaneous frequency – at any f with an acceptable accuracy according to the Bernard's approximation, which is the momentary common reference during the analysis of measured and computed whistlers.

From the $f(t)$ function, constructed by this manner, a $g(t)$ "phase function" can be obtained by integrating equation $f(t)$

$$g(t') = \int_0^{t'} f(t)\mathrm{d}t + \text{constans.} \qquad \textbf{(4.29)}$$

Thus, an approximate whistler waveform used in the matched filter – inside a time window – can be built up rather accurately using the parameters derived from the measured (f,t) dataset. In fact, we proceed in the following way. In the first step using (4.16), we calculate the time t_0 of arrival of a given frequency f_0. At t_0 the approximate waveform function is

$$w(t_0) = A(t_0)\cos(2\pi f_0 t_0 + \alpha), \qquad \textbf{(4.30)}$$

where α is a phase constant and $A(t)$ is the variation of the signal amplitude in time. Then the "whistler" waveform function is at time Δt according to equations (4.29) and (4.30)

$$w(t_0 + \Delta t) = A(t_0 + \Delta t) \cdot \cos\left\{2\pi\left[f_0 t_0 + \int_{t_0}^{t_0+\Delta t} f(t)\mathrm{d}t\right] + \alpha\right\}, \qquad \textbf{(4.31)}$$

where $f(t)$ is found by inverting the *Bernard's* approximation (4.16). It is important to note that the function $f(t)$ and therefore also the reference waveform function $w(t)$ in matched filter has a meaning only when $t > t_n$, where t_n is the travel time of the nose frequency f_n, and $f(t)$ has two values at any $t > t_n$ (one above and one below the nose) in accordance with the inverted relation of (4.16). When computing the waveform this behavior of $f(t)$ had to be taken into account.

For the construction of the matched filter we have to compute the reference waveform with Δf bandwidth around a given frequency f_0. – For the sake of simplicity, in the following, we present the case of $f_0 < f_n$. – The computation is performed as follows: First the time t_0 of the given frequency f_0 is derived from equation (4.16). Then inverting equation (4.16) numerically we determine the frequency f_1 at $t_1 = t_0 + \Delta t$ (where Δt is the sampling interval of the measured whistler data). Assuming that the variation of the instantaneous frequency between f_0 and f_1 can be taken as linear, the value of the integral in equation (4.31) becomes

$$\int_{t_0}^{t_0+\Delta t} f(t)\mathrm{d}t \cong \frac{f_0 + f_1}{2}\Delta t. \tag{4.32}$$

By substituting equation (4.32) into (4.31), the value of the reference (signal-) wave-form function at $t_1 = t_0 + \Delta t$ is

$$w(t_0 + \Delta t) \cong A(t_0 + \Delta t)\cdot\cos\left\{2\pi\left[f_0 t_0 + \frac{f_0 + f_1}{2}\Delta t\right] + \alpha\right\}. \tag{4.33}$$

Repeating the procedure described above for a succession of steps separated from each other by Δt we obtain the waveform at $t_i = t_0 + i\Delta t$ as

$$w(t_i) \cong A(t_i)\cdot\cos\left\{2\pi\left[f_0 t_0 + \left(\frac{f_0 + f_1}{2} + \ldots + \frac{f_{i-1} + f_i}{2}\right)\Delta t\right] + \alpha\right\}. \tag{4.34}$$

This approximate signal form inside a time-widow is valid (usable), if $t_i > t_n$. The relation (4.34) is not valid, if $t_i < t_n$ and this approximation is unstable around the nose $(t_i \cong t_n)$. Approaching the whistler nose it sometimes happened. For computational reasons, inverting equation (4.16) we obtain not the appropriate but another value of frequency f above the nose. In that case, the computation was repeated with another starting value until a "proper" root was obtained below the nose-frequency.

The reference waveform above the nose-frequency can be computed in a similar way. If the Δf frequency band and the f_0 mean frequency has been chosen in a way that it includes the nose frequency, the computed signal should be the addition of two signals: one under the nose frequency, the other is above it.

The cause of the computational problems around the nose is the construction of the reference "whistler" signal itself and this instability will disappear using the real computed transfer function (see Chapters 1 and 2) as reference signal. In this case, the analysis of measured signals is solved, however, the simultaneous investigation of measured and computed whistlers is not so demonstrative, as in applying for these comparative investigations another common reference, the (4.34) approximation. Naturally, the (4.34) approximation is not accurate not only around the nose, therefore systematic differences will appear during the analysis of measured signals or computed by the full wave solution, which are similar in both cases.

4.2 The fine structure of signals

To demonstrate the potential advantage of the matched filtering in wave analysis, let us present the results of matched filtering of a measured whistler signal, because the analysis of the fine structure of signals could be very informative. The data were recorded at Halley Station of British Antarctic Survey in Antarctica (76°S 27°W, L ≅ 4.3), on 21 July 1984, beginning at 17:14:14 UT, using two vertical loop aerials aligned in north-south and east-west directions. The sampling rate was 10 kHz per component and the length of sample 3.27 s, of which the first part was discarded for the purposes of this presentation.

The result of conventional FFT spectral analysis is shown in Fig. 4.1. The spectrogram shows two strong whistlers (traces A and B) with some "triggered emissions" above the upper cut-off frequencies. A broad hiss band can also be seen at around 2 kHz with a bandwidth about 600~800 Hz.

To demonstrate the usefulness of matched filtering in the analysis of whistlers, it is sufficient to present the results for only one component of the wave field. We have used here the-east-west data, noting that the north-south data yielded similar results.

4.2.1 Matched filtering of the data

Let us scale the trace "A" from the dynamic spectra, and calculate the best fit parameters by the FIT method. The obtained parameters $(f_{HE} = 8632\,\text{Hz}, D_0 = 84.6\,\text{s}^{-1/2}, T = 1.968\,\text{s})$ were used to construct the matched filter. The matched filter were constructed around the instantaneous frequency f_m as it was described in the section 4.1.2.2. According to *Hamar and Tarcsai* (1982) the filter output can be approximated as

$$s_0(t) = A \operatorname{sinc}[\Delta f(t - t_m)] \cos[2\pi f_m(t - t_m)], \tag{4.35}$$

where t_m is the time of arrival of the instantaneous frequency f_m and A is proportional to the average amplitude of the whistler around f_m. Using the Hilbert transform we can eliminate the high frequency cosine function and obtained the envelope, the maximum of the envelope curve is at $t = t_m$. A smoothing of the final output signal is possible using a second filtering in which the reference (filter) function is constructed by the theoretical envelope curve of function (4.35).

Testing the procedure with filters of different bandwidth and with different f_m frequencies, the filter bandwidth of 200 Hz proved to be the most convenient in the case of whistler signals measured e.g. at Halley Station of British Antarctic Survey. The application of filter functions with smaller bandwidth gave poor resolution in time, i.e. poor separations of closely spaced whistler traces. Another disadvantage of the smaller bandwidth is the less suppression of the output noise and increasing in uncertainty of determining t_m (4.11). However, the application of wider bandwidth produces longer reference functions in time and therefore the matched filter is more sensitive to small differences between the dispersions of the filter function and the measured (or computed) whistler signal. According to computer simulations, the cca. 200 Hz

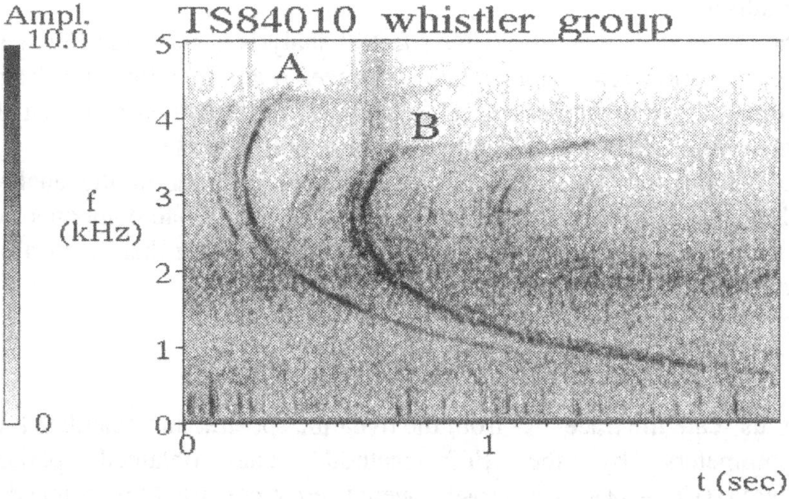

Figure 4.1. The dynamic spectrum of two whistlers (traces A and B) recorded digitally at Halley, Antarctica on 21 July 1984. The time axis starts at 17:14:14 UT. - Courtesy of dr. A.J. Smith, British Antarctic Survey, Cambridge, U.K

filter bandwidth enables even the visual identification of whistler components if they are separated in time at least by $1.4/\Delta f$, i.e. by 7 ms. The procedure of matched filtering outlined here was demonstrated using a reference function as input signal, which was constructed by the fitted parameters of trace A in Fig. 4.1. The results are shown in Fig 4.2. The input signal is the waveform around the frequency $f_m = 1850$ Hz.

The same procedures were performed when the input signal was the measured whistler trace A. The main steps and the results of filtering at $f_m = 1850$ Hz are shown in Fig. 4.3. Comparing Fig. 4.3 to Fig. 4.2, it is clear that the filter outputs of the test signal and the measured one are quite similar. The secondary maxima of the smoothed output of the measured signal show noise or other (small) signals than the main trace. The filter output, the envelope, and the smoothed envelope, represent the main stages of the procedure. The results of this procedure are the arrival time and the magnitude of the local maxima of the smoothed envelope as the local amplitude of the analysed signal at the given frequency (f_m).

The procedure was repeated in this case using 5 Hz frequency steps in f_m across the whole range of FFT pattern of the measured whistler (Fig. 4.1 - trace A). The 1% value of the largest maximum of smoothed envelope-output of a given (measured or computed) whistler signal is used as an output threshold, eliminating the effects of measured (or simulated) random noise and the small resonance of matched filter, which are methodological artefacts.

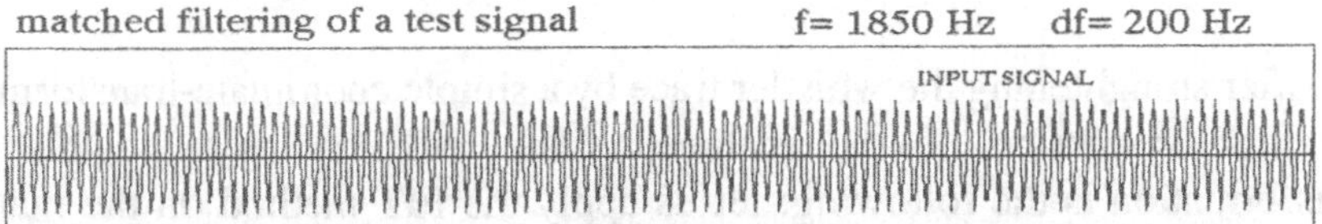

Figure 4.2. Determining the arrival time of the instantaneous frequency 1850 Hz of an input reference signal computed by the fitted parameters of whistler trace A in Fig. 4.1

At the end of this procedure we obtained a set of (f,t) pairs with the corresponding amplitude values. If the (f,t) pairs are plotted and the amplitude of the pairs is scaled e.g. with a greyscale palette, the result is the MF dynamic spectrum (Fig. 4.4). If we compare it with the conventional dynamic spectrum (see in Fig. 4.1) the noise suppression and the more sharp pattern of the trace is apparent. (The diffuse character of the filtered signal at the nose is artefact. The cause of this is the very approximate character of the reference signal, which was discussed in connection of relation (4.34).)

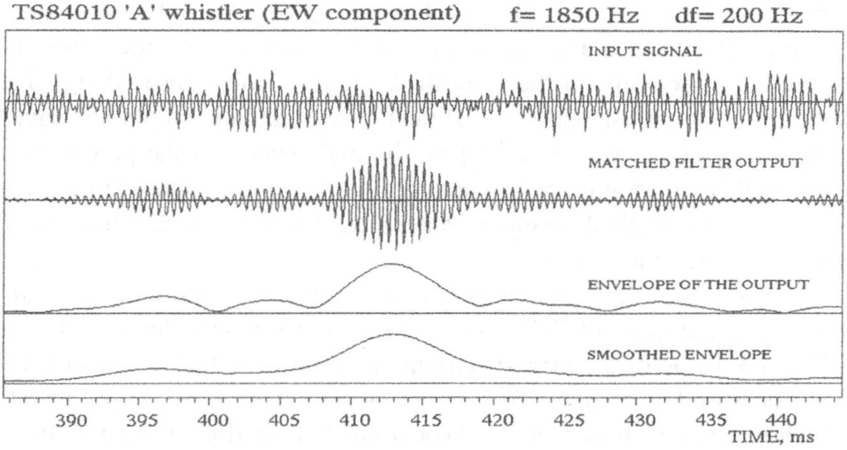

Figure 4.3. Determining the arrival time of the instantaneous frequency 1850 Hz of the measured whistler trace A in Fig. 4.1. The moment of arrival is the time data of the maximum of the smoothed envelope curve

After straightening the whistler trace by a simple coordinate-transformation the high resolution of the matched filter (MF) is more visible. The transformation is the following: let us apply the FIT method on the matched filtered dynamic spectrum and subtract the travel time data of this fitted *f-t* curve (Bernard's approximation) from the $t(f)$ data of the analysed signal ("measured data"). Let the remained time difference be "dt_{TR}" in horizontal axis. A curve with Bernard's approximation (4.16) is represented by a straight vertical line on such transformed plots, while spherics would appear hyperbolic, like the isochrones in Fig. 4.5. Fig 4.5 presents the transformed plot of matched filtered trace A signal presented in Fig. 4.1.

The high accuracy of the obtained *f-t* pairs makes it possible to "zoom out" the transformed plot – Fig. 4.6. The increased part of the dynamic spectrum made it possible to observe a special effect around $f = 3.8$ kHz. The single trace split into two traces, separated by 7 ms. This effect can be due to the rapid decreasing of the whistler amplitude which may be caused by effects during the propagation (e.g. interference), or other effects – see part 4.4.1. In this case – as the model calculations confirmed – the interference between the close components or the actual form of the source signal (excitation) can result in similar effects. These small deviations can be called the fine structure of a whistler. As can be seen in Fig 4.5 and Fig.4.6, the transformed trace locally deviates more or less from the vertical. These deviations in time are known as travel time residuals (e.g. *Tarcsai et al.*, 1989).

During the fitting (FIT-method) the Bernard's approximation to the final MF output of the measured (or computed) whistler signal, we can use more accurate

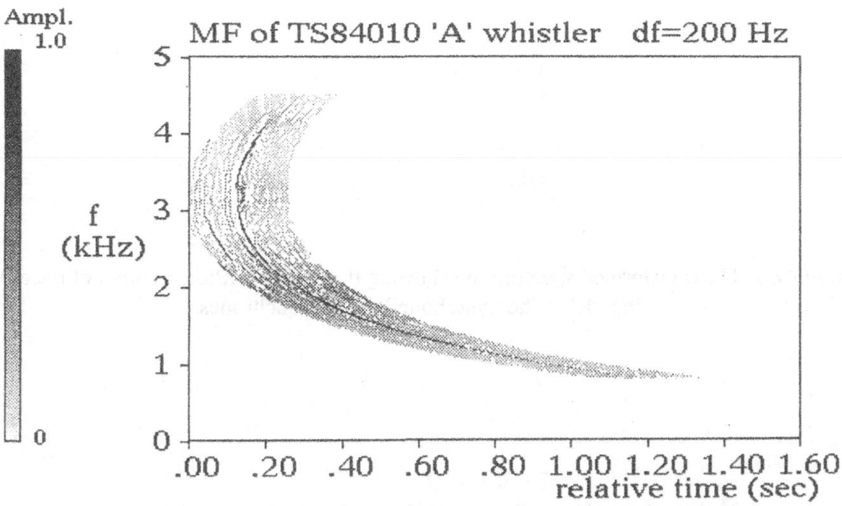

Figure 4.4. The matched filtered dynamic spectrum of the whistler trace A in Fig. 4.1.

input *f-t* pairs because the MF output is a sharp line if the original signal has the same character. The accurate input data decrease the standard deviations of the fitted (estimated) parameters. In our example:

f_{HE} = 8548 ± 3 Hz, L = 4.675 ± 0.0,
f_0 = 3174 ± 1 Hz, n_{eq} = 177 ± 0 cm^{-3},
D_0 = 87.8 ± 0.1 $s^{-1/2}$, n_1 = 2700 ± 3.1 cm^{-3},
T = 2.068 ± 0.002 s, n_t = (3.12 ± 0.004)$\cdot 10^{13}$cm^{-2};

where n_{eq} is the equatorial electron density at distance L, n_1 is the electron density of high atmosphere at 1000 km height above the Earth's surface and n_t is the total electron content along the propagation path of the whistler signal (i.e. "tube"-content). Based on these parameters the computed travel time residuals of the investigated whistler are presented in Fig. 4.7.

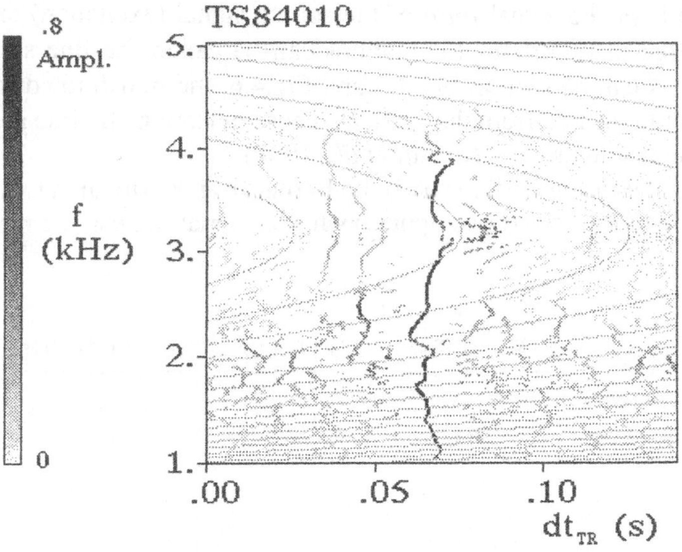

Figure 4.5. The transformed spectrogram showing the matched filtered output of trace A in Fig. 4.1. (The hyperbolic lines are isochrones.)

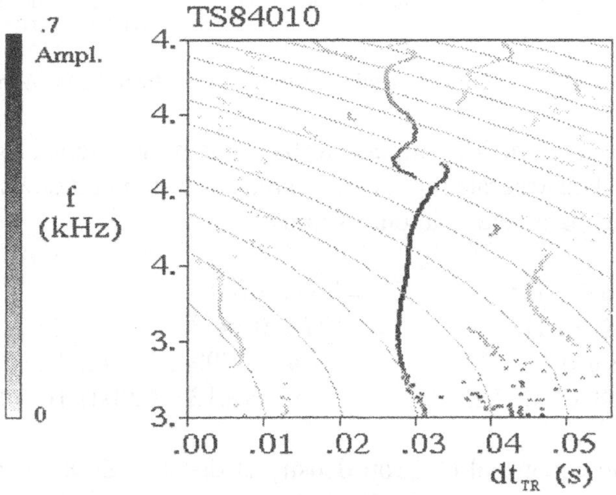

Figure 4.6. A "zoomed" part of MF trace A around 3800 Hz

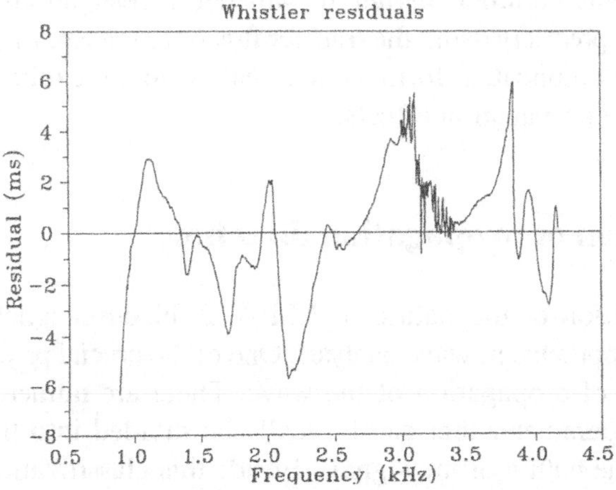

Figure 4.7. Travel time residuals of the measured whistler trace A (see in Fig. 4.1). (The instability around the nose ($f \cong$ 3174 Hz) is artefact – see in part 4.1.2.2.)

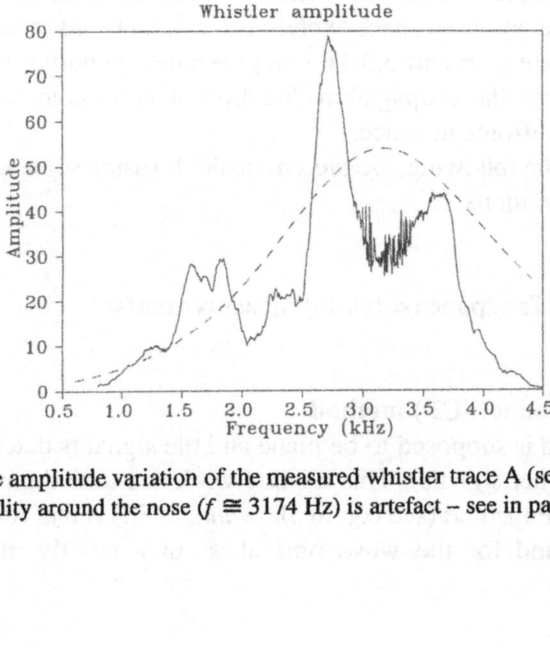

Figure 4.8. The amplitude variation of the measured whistler trace A (see in Fig. 4.1). (The instability around the nose ($f \cong$ 3174 Hz) is artefact – see in part 4.1.2.2.)

The amplitude variation of the main trace of whistler A is shown in Fig. 4.8. It seems to be more complicated than might be expected by making the simple assumption that the energy per unit bandwidth is constant (dotted line in Fig. 4.8) – which was common earlier in whistler investigations and it can be computed more precisely using the transfer function, see e.g. Fig. 2.7 –, and the cause of this complicated form is the real form of excitation and partly (probably) some propagation effects.

4.3 Derivation of propagation direction

The recognition of the nature of ELF-VLF electromagnetic waves is of fundamental importance in wave analysis. One of the crucial physical parameters is the direction of propagation of the wave. There are numerous methods to determine this parameter that can basically be divided into three categories, depending on the nature of the signals. Beside this classification, the methods also differ depending on whether they are applied to data measured in space or on the ground.

The basic theoretical problems are about the correct definition of the direction of propagation and of the velocity of the propagation of electromagnetic energy in monochromatic and non-monochromatic cases which were solved earlier (*Ferencz*, 1977b, 1978b, 1987, 1996). However, the instrumentation of the existing satellite and ground measurements is not enough to use a highly sophisticated theoretical analysis of these phenomena. The realisation of some proposed space experiment, e.g. on board of the International Space Station (see e.g. in part 3.3.1), will give a new opportunity to analyse in a more precise way the propagation direction of non-monochromatic waves having broad wavefronts in space.

Therefore, in the following, we present methods (using some simplifications) for practical applications.

4.3.1 Methods for space (satellite) measurements

4.3.1.1 Cross-product (CP) method

The wave field is supposed to be plane and the signal is deterministic (e.g. whistlers, transmitter signals). This method works on time domain. The basic principle of the CP method (see e.g. in *Shawhan*, 1983) is that for the magnetic field vector $\overline{B}$ and for the wave normal $\overline{k}$ in a strictly monochromatic approximation

$$\bar{k}\cdot\bar{B}=0, \tag{4.36}$$

as a well known consequence of Maxwell's equations (1.5a). Namely, if $\bar{E}=\bar{E}_0\exp(j\omega t-\bar{k}\bar{r})$ and $\bar{B}=\bar{B}_0\exp(j\omega t-\bar{k}\bar{r})$ and

$$\begin{aligned} &\bar{\nabla}\times\bar{E}=-\frac{\partial\bar{B}}{\partial t}, && \bar{k}\times\bar{E}=\omega\bar{B},\\ & && \Rightarrow \\ &\bar{\nabla}\cdot\bar{B}=0, && \bar{k}\cdot\bar{B}=0. \end{aligned} \tag{4.37}$$

If the wave is non-linearly polarized e.g. rotating (see relation (1.39) or others in earlier parts), which is true practically for all waves propagating in magnetized plasma, than

$$\bar{K}=\bar{B}(t_1)\times\bar{B}(t_2), \qquad t_1\neq t_2, \tag{4.38}$$

where $\bar{K}$ is a vector parallel to wave normal $\bar{k}$. Computing the θ and ϕ polar angles (polar coordinates) of the vector $\bar{K}$ (see Fig. 4.9) let us define the vector $\bar{e}_k$ as

$$\bar{e}_k=\frac{\bar{K}}{|\bar{K}|}, \tag{4.39}$$

and we have the direction of propagation of the wave in a frame of reference (Fig. 4.9). In the case of a real, noisy signal $\bar{K}$ should be computed for several different pair of (t_1,t_2) and the average is taken. The t_1 and t_2 should be chose that the angle between $\bar{B}(t_1)$ and $\bar{B}(t_2)$ be close to 90 degrees to reduce the instability. There is an ambiguity in the sign of the wave normal vector, i.e. that the real direction of propagation is $\bar{e}_k$ or $(-\bar{e}_k)$, however this can be removed by using electric field components and calculating the Poynting vector $\bar{S}(t)=\bar{E}\times\bar{H}$.

4.3.1.2 Means method

In this case,(*Means*, 1972) the wave field is supposed to be monochromatic and plane or formed by a finite number of plane waves, but the signal is non-deterministic (complex whistler traces, transmitter signals and discrete emissions). This method works on frequency domain approximation.

Let $\bar{G}$ be the general field vector derived from electric $\bar{E}$ and magnetic $\bar{B}$ vector, $\bar{G}=(\bar{E},\bar{B})$. The spectral matrix

$$\bar{s}(\omega)=\bar{G}(\omega)\circ\bar{G}^*(\omega), \tag{4.40a}$$

where $\circ$ denotes the dyadic product, the $*$ denotes the complex conjugate. The spectral matrix is Hermitian matrix, i.e. the real part is symmetric and the imaginary part is antisymmetric. In components

$$s_{ij}(\omega) = \overline{G}_i(\omega)\overline{G}_j^*(\omega), \qquad i = 1,...,6 \text{ and } j = 1,...,6 \tag{4.40b}$$

and it is clear from equation (4.38) that the elements of s_{ij} are the autocorrelation and cross correlation spectra of field components. Let us consider the lower-right submatrix of the spectral matrix, i.e. the spectral matrix is derived from magnetic field only. If a monochromatic plane wave propagates along the axis z with $\theta = 0$ (see Fig. 4.9), the wave components can be written as

$$\begin{aligned} B_x(t) &= B_{x0} \cdot e^{j(\omega t + \varphi_x)}, \\ B_y(t) &= B_{y0} \cdot e^{j(\omega t + \varphi_y)}, \\ B_z(t) &= 0, \end{aligned} \tag{4.41}$$

and the spectral matrix is

$$\begin{vmatrix} B_{x0}^2 & jB_{x0}B_{y0} & 0 \\ -jB_{x0}B_{y0} & B_{y0}^2 & 0 \\ 0 & 0 & 0 \end{vmatrix} = \begin{vmatrix} B_{x0}^2 & 0 & 0 \\ 0 & B_{y0}^2 & 0 \\ 0 & 0 & 0 \end{vmatrix} + jB_{x0}B_{y0}\begin{vmatrix} 0 & 1 & 0 \\ -1 & 0 & 0 \\ 0 & 0 & 0 \end{vmatrix}. \tag{4.42}$$

Let us apply a general rotation to the imaginary part of (4.42) with Eulerian angles $\psi, \theta, \phi,$, it gives

$$\overline{\overline{s}}_{ij}' = B_{x0}B_{y0}\begin{vmatrix} 0 & \cos\theta & -\sin\theta\cos\phi \\ -\cos\theta & 0 & \sin\theta\sin\phi \\ \sin\theta\cos\phi & -\sin\theta\sin\phi & 0 \end{vmatrix}. \tag{4.43}$$

Here, we got the polar angles of the wave normal vector, if the wave propagates in a direction other than the axis z. Comparing (4.43) with (4.40b),

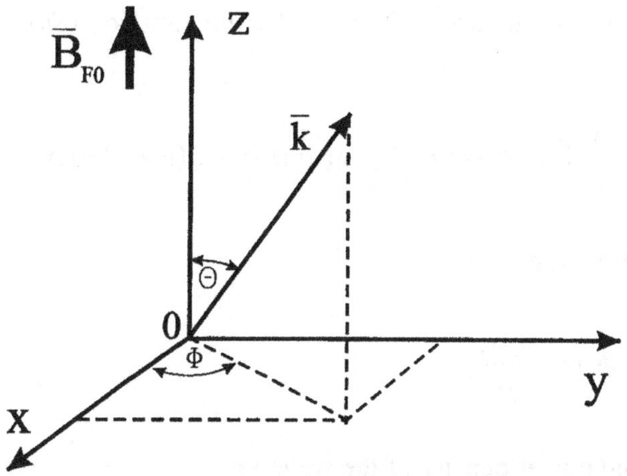

Figure 4.9. Definition of polar co-ordinates θ,ϕ for computation of wave normal $\bar{k}$, as the frame of reference. The Earth's magnetic filed B_{Fo} is parallel to axis z. (Principal axis system.)

$$\begin{aligned} \eta_x &= \operatorname{Im}[s_{56}(\omega)], \\ \eta_y &= -\operatorname{Im}[s_{46}(\omega)], \\ \eta_z &= \operatorname{Im}[s_{45}(\omega)]. \end{aligned} \tag{4.44}$$

From this vector, the polar angles of the wave normal $\bar{k}$ can be calculated as follows

$$\begin{aligned} |\cos\theta| &= \eta_z / |\bar{\eta}|, \\ \operatorname{tg}\phi &= \eta_z / \eta_x. \end{aligned} \tag{4.45}$$

Equation (4.41) corresponds to equation (4.38) in frequency domain, thus in the case of deterministic waves, the cross-product method and the means method are equivalent.

4.3.1.3 The Wave Distribution Function (WDF)

The wave field is supposed to be random and the signal is non-deterministic (e.g. in the case of chorus, hiss). The WDF method (*Storey and Lefeuvre*, 1979, 1980) works on frequency domain and use a monochromatic-type description of the problem. In the case of superposition of several non-deterministic signals the wave normal $\bar{k}$ is meaningless. Instead of $\bar{k}$, the $g(\omega,\theta,\phi)$ wave (energy) distribution function (WDF) can be defined and it specifies how the wave energy is distributed relatively to the wave normal $\bar{k}$ as a function of frequency. The

WDF is related to the spectral matrix by the equation (*Storey and Lefeuvre,* 1979, 1980)

$$s_{ij}(\omega)=\frac{\pi}{2}\int_{-1}^{1}\int_{0}^{2\pi} a_{ij}(\omega,\theta,\phi,)\cdot g(\omega,\theta,\phi,)\cdot \mathrm{d}(\cos\theta)\cdot \mathrm{d}\,\phi. \qquad \textbf{(4.46)}$$

The kernel function a_{ij} is

$$a_{ij}(\omega,\theta,\phi,)=\frac{G_i G_j^*}{\rho}, \qquad \textbf{(4.47)}$$

where ρ is the power density of the wave i.e.

$$\rho=\frac{1}{2}\left(\varepsilon E^2+\mu H^2\right)$$

for a monochromatic wave in isotropic, linear media, where ε is the permittivity and μ is the permeability of the medium. The kernel function a_{ij} contains the plasma characteristics because in relation (4.47) the G_i components (see in point 4.3.1.2) mean the $\overline{E}$ and $\overline{B}$ components of the propagating (monochromatic) wave. As is known – see e.g. in the Appendix B – the relation of the $\overline{E}$ and $\overline{B}$ components depend on the plasma parameters. An example of the result of WDF method applied to GEOS1 data can be seen in Figs. 4.10-4.11. In the case of plane waves and deterministic signals, the WDF method equivalent to Means method.

There is another similar method to WDF method, called Field Distribution Function method (*Sonwalkar and Inan,* 1986, 1988), here the distribution function describes the distribution of wave amplitude and phase instead of wave energy.

4.3.1.4 The Matched Filtering and Parameter Estimation (MFPE) method

The previously described methods are general methods, i.e. in the case of deterministic signals they suppose nothing about the wave itself. Though in the case of deterministic signal, usually we know the signal. Using this a priori information the wave normal vector $\overline{k}$ can be determined with much higher accuracy (*Lichtenberger et al,* 1987) in the case of real, noisy signals (and applying monochromatic-type modelling). The method combines the a priori information through matched filtering and enhances the accuracy by statistical parameter estimation. These techniques are described in detail in part 4.2. The wave field is supposed to be plane or formed by several plane waves and the signal is deterministic.

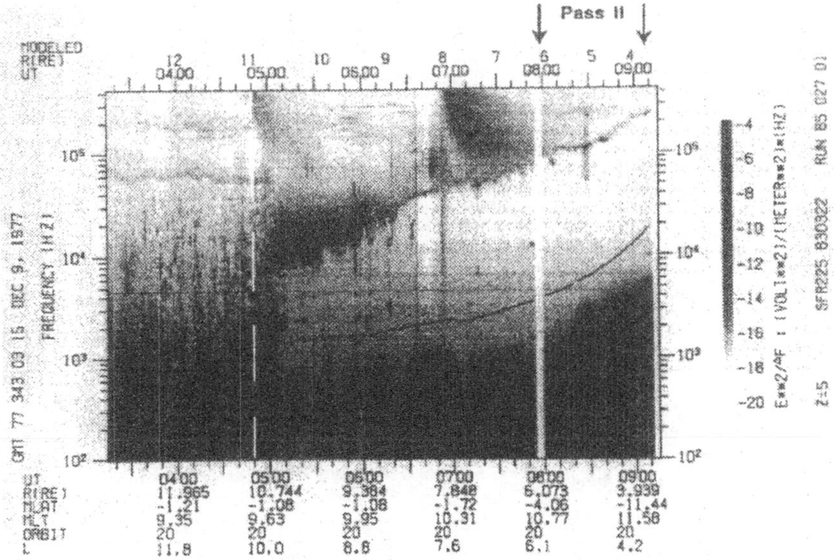

Figure 4.10. Spectrogram of hisses observed by ISEE1 satellite on December 9, 1977 (*Storey et al.*, 1991, Fig.2.)

The time variation of waves propagating in a cold, collisionless plasma can be written in the form

$$\begin{aligned} E_x &= E_{x0}\cos(\omega t+\alpha), & B_x &= B_{x0}\cos(\omega t+\phi), \\ E_y &= E_{y0}\cos(\omega t+\beta), & B_y &= B_{y0}\cos(\omega t+\chi), \\ E_z &= E_{z0}\cos(\omega t+\gamma), & B_z &= B_{z0}\cos(\omega t+\psi), \end{aligned} \tag{4.48}$$

where the electric field $\overline{E}$ is perpendicular to the magnetic field $\overline{B}$, i.e. $\overline{E}\cdot\overline{B}=0$. From this it follows that if we know five field components, the sixth one can be calculated. On the other hand, there is no need to know the absolute values of the five amplitudes and phases. As is easy to see, it is sufficient to know the ratios of the amplitudes and the phase differences. In fact, the amplitude of the filter output is determined by the input amplitude up to a multiplicative constant, the output phase up to an additive constant. Thus, the sixth component can be calculated according to the previous statement. In this MFPE method, the amplitudes and phase constants in equation (4.48) are taken from the matched filtering and the parameter estimation procedure shown in part 4.2.

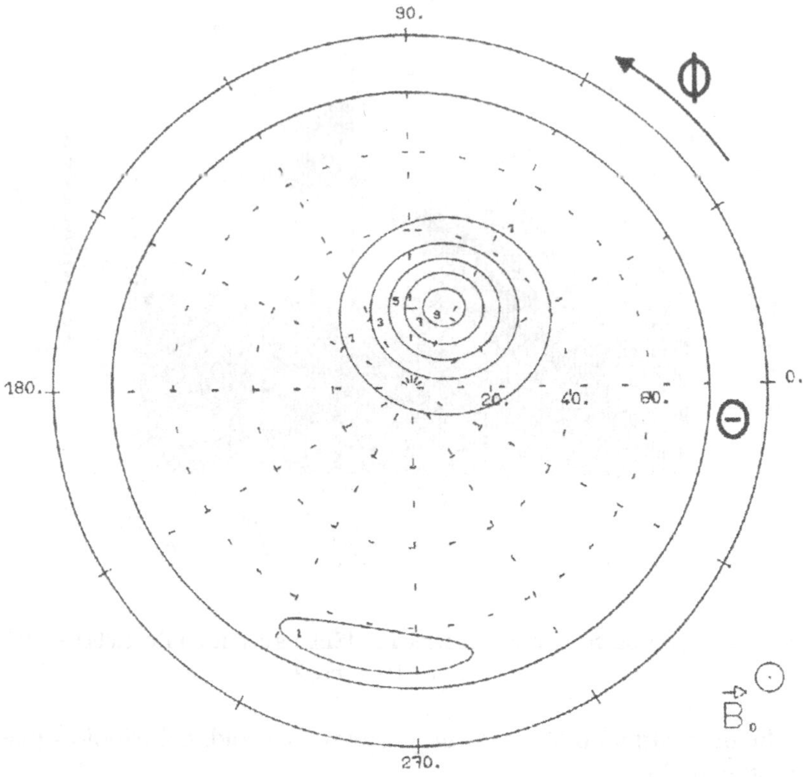

Figure 4.11. WDF of hisses presented in Fig. 4.10. at 1494 Hz and 08:24:32 UT (*Storey et al.*,1991, Fig.6.)

Since $\overline{B} \cdot \overline{k} = 0$, a vector parallel to the vector $\overline{k}$ can be constructed by

$$\overline{\kappa} = \overline{\hat{B}}(t_1) \times \overline{\hat{B}}(t_2), \qquad t_1 \neq t_2, \tag{4.49}$$

where the caret superscript denotes the magnetic field vector computed by estimated amplitude ratios and phase differences. For the case of simplicity we take $t_1 = 0$ and $t_2 = \pi / 2$

$$\begin{aligned} B_x(0) &= B_{x0}\cos(\phi), & B_x(\pi/2) &= B_{x0}\sin(\phi), \\ B_y(0) &= B_{y0}\cos(\chi), & B_y(\pi/2) &= B_{y0}\sin(\chi), \\ B_z(0) &= B_{z0}\cos(\psi), & B_z(\pi/2) &= B_{z0}\sin(\psi). \end{aligned} \tag{4.50}$$

Even if this looks the same as the CP method (part 4.3.1.1), it differs from that where the wave normal direction is computed directly from the measured noisy signal detected by the magnetic aerials. From the vector product (4.49), we only know the direction of $\bar{k}$, but not the sense of $\bar{k}$. However, the sense of $\bar{k}$ is identical with that of the sense of the Poynting vector $\bar{S}$ which can be computed as

$$\bar{S} = \frac{1}{\mu_0} \bar{E} \times \bar{B}. \tag{4.51}$$

Thus, by the aid of $\bar{S}$, we can fix the sense of $\bar{e}_k$ (i.e. the sense of $\bar{\kappa}$). Let be

$$\bar{e}_k = \frac{\operatorname{sgn}(\bar{\kappa} \cdot \bar{S}) \cdot \bar{\kappa}}{|\bar{\kappa}|} \tag{4.52}$$

and from this, the vector $\bar{k} = \kappa \bar{e}_k$ using this monochromatic-type description, where the definition of $\operatorname{sgn}(x)$ is

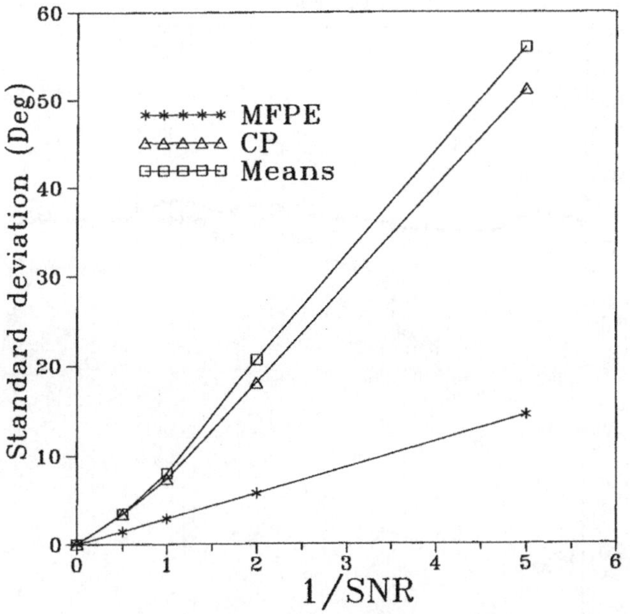

Figure 4.12. Comparison of the CP, Means and MFPE methods for different signal to noise ratios (SNR)

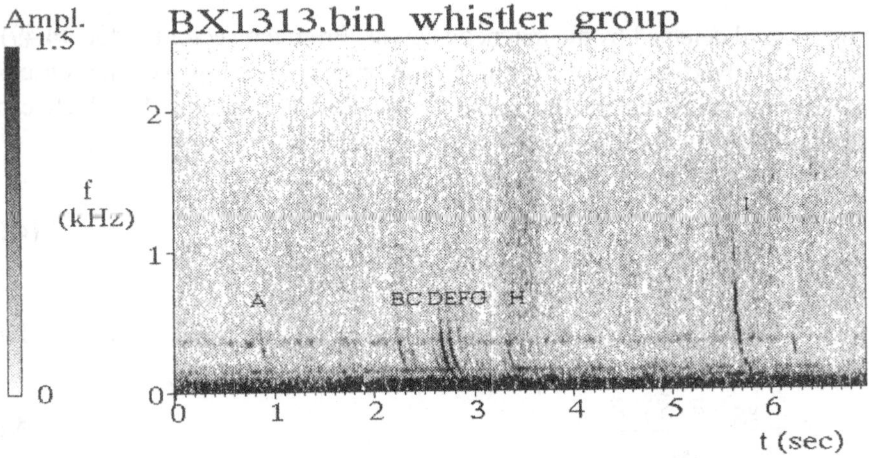

Figure 4.13. Upgoing (i.e. fractional hop) whistlers recorded on board AUREOL-3 satellite on 16 March 1982, 04:47:17 UT

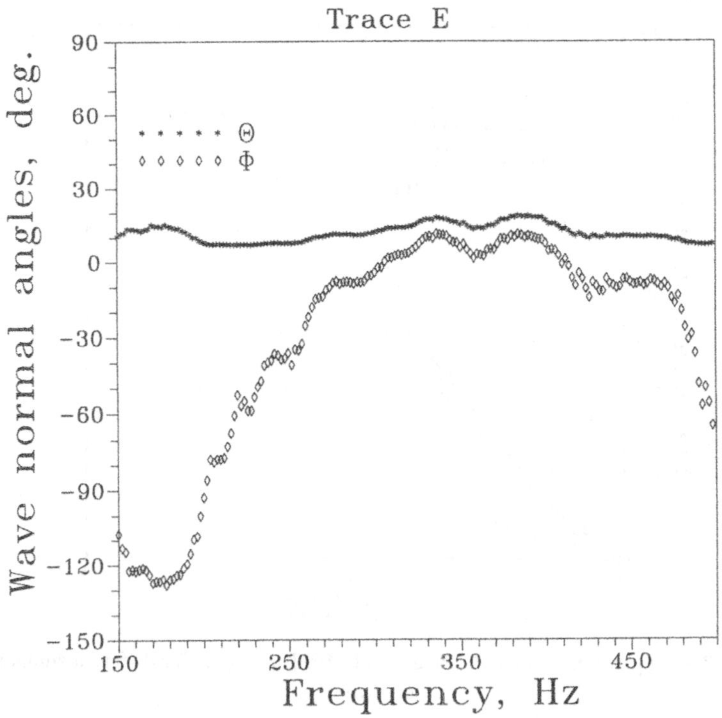

Figure 4.14. Frequency dependence of θ and ϕ wave normal angles for trace *E* in Fig. 4.13

$$\operatorname{sgn}(x)=\begin{cases}1, & \text{if} \quad x>0,\\ 0, & \text{if} \quad x=0,\\ -1, & \text{if} \quad x<0.\end{cases} \tag{4.53}$$

To sum up, the MFPE method uses the three estimated magnetic field components to determine the direction of wave normal vector $\bar{k}$ and uses the estimated electric field components together with the magnetic components, only to determine the sense of $\bar{k}$.

A comparative study including CP, Means and MFPE methods had been done, and proved the superiority of MFPE method to others – see Fig. 4.12. Fig. 4.13 shows the dynamic spectrum of whistlers recorded on board AUREOL-3, the frequency-dependence of wave normal angles can be seen on Fig. 4.14. The SNR of this data set is typical for satellite measurements. It is clear that the upgoing whistler trace in this measurements propagated with a wave normal vector rotating around the Earth's magnetic field, thus probably in a guiding inhomogeneity.

4.3.2 Methods for ground based measurements

As is described in chapter 3.1, on ground based measurements only two horizontal magnetic and one vertical electric field components can be measured. This yields a series of simplifications and approximations in the computation of direction of arrival (i.e. the final direction of propagation) using monochromatic-type modelling.

4.3.2.1 The goniometer method

The wave is supposed to be plane, linearly polarized and propagates horizontally. Thus, the magnetic field measured with two orthogonal loops, the induced signals in the loops of a given frequency

$$U_x = AB\cos\phi\cdot\cos\omega t,$$
$$U_y = AB\sin\phi\cdot\cos\omega t,$$

where B is the magnetic induction of the signal coming from the direction ϕ and A is a constant.

Multiplying U_x and U_y by $U_0\sin\omega_g$ and $U_0\cos\omega_g t$ respectively and adding them

$$U_g = U_x + U_y = A\cdot B\cdot U_0\cdot\cos\omega t\sin(\phi+\omega_g t). \tag{4.54}$$

Here ω_g is a fixed frequency and $\omega_g \ll \omega$. This is the classical rotating loop radio direction finder. The bearing can be determined by finding the minima and maxima of U_g. (See more in *Bullough and Sagredo,* 1973; *Sagredo and Bullough,* 1973.)

4.3.2.2 Methods for elliptically polarized waves

However, according to the measurements (*Okada et al.*, 1977) the major part of waves on the ground are neither linearly polarised nor propagate horizontally, i.e. the real waves are elliptically polarized and propagate with a zenith angle other than 90 degrees. Thus the e.m. field observed on the conductive ground can be written in a complex form as

$$
\begin{aligned}
E_z &= -A\sin\theta e^{j\omega t}, \\
B_x &= A\sin\phi e^{j\omega t} - B\cos\theta\cos\phi e^{j(\omega t-\alpha)}, \\
B_x &= -A\cos\phi e^{j\omega t} - B\cos\theta\sin\phi e^{j(\omega t+\alpha)}.
\end{aligned}
\qquad (4.55)
$$

The equation (4.55) describes an elliptically polarized wave and, as is well known, an elliptically polarized wave can be described by two linearly polarised waves propagating with the wave normal described by zenith angle θ and azimuth angle ϕ. The electric field of the first linearly polarized wave is perpendicular to the plane of incidence and the electric field of the second one is in the plane of incidence. The phase difference between these two waves is α. The field components are coupled by the plane wave solution of the

$$
\overline{\nabla} \times \overline{B} = \varepsilon_0 \mu_0 \frac{\partial \overline{E}}{\partial t}
$$

Maxwell's equation in the most simple monochromatic approximation, i.e. if $\overline{G} = \overline{G_0} \exp j(\omega t - \overline{k}\overline{r})$, than

$$
\begin{aligned}
&\overline{k} \times \overline{B} = -\frac{k}{c}\overline{E}, \quad \text{and} \\
&E_z = -e_{kx}B_y + e_{ky}B_x,
\end{aligned}
\qquad (4.56)
$$

where $\overline{e}_k$ is a unit vector parallel to the wave normal $\overline{k}$. Depending on which combination of the three complex amplitudes in (4.55) and (4.56) is used, the formulas of azimuth and zenith angles are the following

a) in goniometer method (*Bullough and Sagredo,* 1973)

$$\mathrm{tg}\,2\phi = \frac{2\cdot \mathrm{Re}(B_x B_y^*)}{B_y B_y^* - B_x B_x^*}, \quad \text{and } \theta \text{ is not determined;} \tag{4.57}$$

b) tracking DF method (*Leavitt,* 1975)

$$\mathrm{tg}\,\phi = \frac{\mathrm{Re}(-E_z B_x^*)}{\mathrm{Re}(E_z B_y^*)} \quad \text{and}$$

$$\sin\theta = \frac{\left\{\left[\mathrm{Re}(E_z B_y^*)\right]^2 + \left[\mathrm{Re}(E_z B_x^*)\right]^2\right\}^{1/2}}{B_x B_x^* + B_y B_y^*} \quad \text{only for large } \theta \text{ values;} \tag{4.58}$$

c) in No Polarisation Error (NPE) method (*Tsuruda and Hayashi,* 1975)

$$\mathrm{tg}\,\phi = \frac{\mathrm{Im}(-E_z B_y^*)}{\mathrm{Im}(E_z B_x^*)} \quad \text{and}$$

$$\sin\theta = \frac{\left\{\left[\mathrm{Im}(E_z B_x^*)\right]^2 + \left[\mathrm{Im}(E_z B_y^*)\right]^2\right\}^{1/2}}{\mathrm{Im}(B_x B_y^*)} \quad \text{for all } \theta \text{ values.} \tag{4.59}$$

Both goniometer and tracking-DF methods have azimuth error when $\theta < 90$ degrees, while NPE gives the accurate azimuthal bearing for any θ and provides the θ itself as well. However, in addition to the fact that the waves are generally elliptically polarized and propagate with a zenith angle other than 90 degrees – a situation that can be handled perfectly by NPE method – the waves usually follow multiple ray paths between the source and receiver in the ionosphere-Earth wave-guide (Fig. 4.15). Detailed analyses of the two effects (polarisation and multipath propagation) shows that both yield systematic errors (*Strangeways,* 1980), that can be as high as 90 degrees in the case of azimuthal bearing (Figs. 4.16) even for methods free of polarisation errors (NPE). However, the two systematic error sources tend to cancel each other giving a combined error which is smaller than either error source alone (Fig. 4.17). Thus the usage of vertical electric field gives worse result than the usage of the two horizontal magnetic components alone. *Strangeways and Rycroft* (1980) shows that the total bearing error can be significantly reduced by averaging over a 2 kHz bandwidth. However, at the moment there is a lack of good experimental evidence to support model calculations, probably because of the inadequate frequency resolution of the generally used direction finding methods.

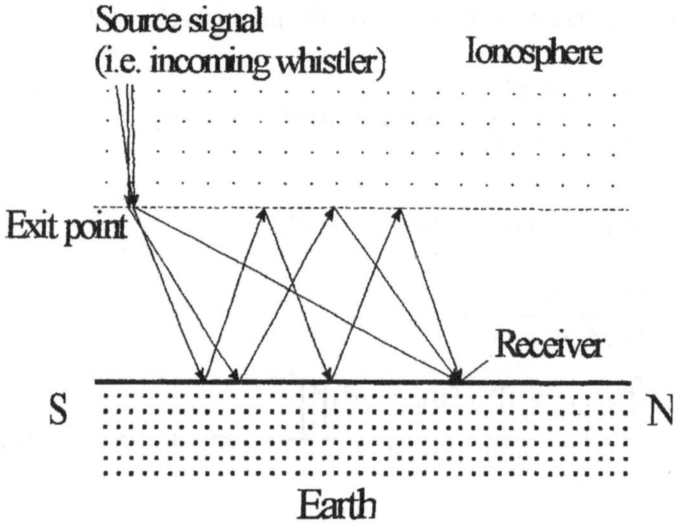

Figure 4.15. Multiple ray paths between exit-point and receiver (neglecting the real waveguide-mode propagation pattern in the Earth-ionosphere waveguide)

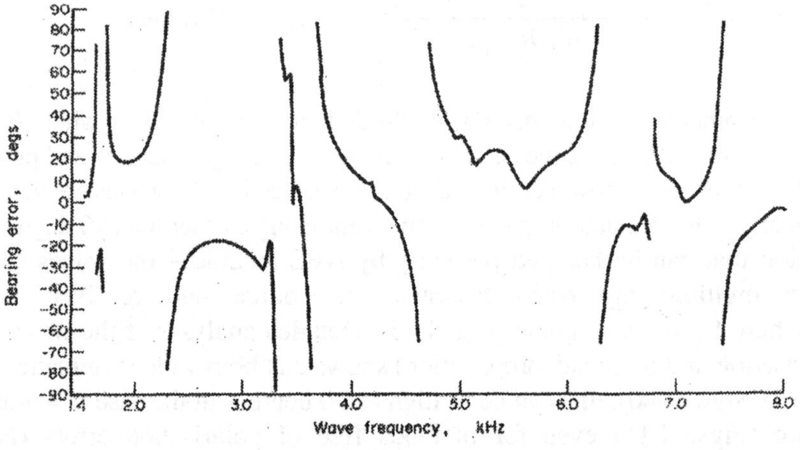

Figure 4.16a. Bearing error due to multipath propagation in the Earth-ionosphere waveguide for the crossed loops goniometer for a propagation distance of 500 km. Summer night ionosphere and poorly conducting ground. (*Strangeways,* 1980, Fig. 8.)

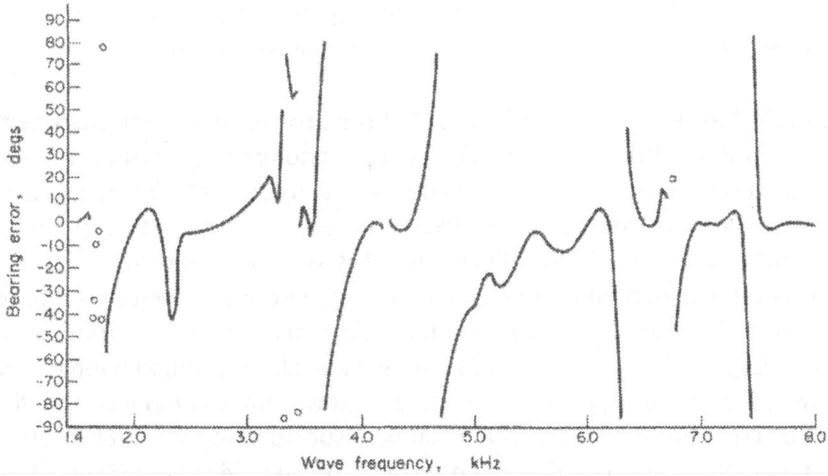

Figure 4.16b. Bearing error due to polarisation for the crossed loops goniometer for a propagation distance of 500 km. Summer night ionosphere and poorly conducting ground. (*Strangeways*, 1980, Fig. 9.)

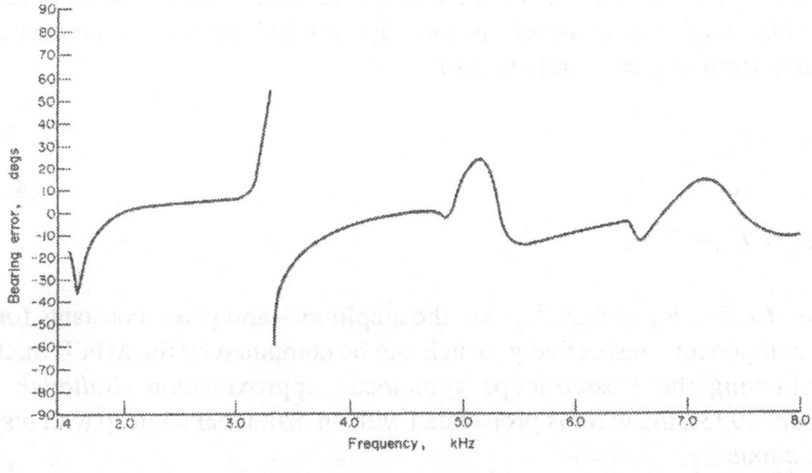

Figure 4.17. Combined systematic bearing error for the crossed loops goniometer for a propagation distance of 150 km. Summer night ionosphere and poorly conducting ground. (*Strangeways*, 1980, Fig. 11.)

The MFPE (**M**atched **F**iltering and **P**arameter **E**stimation) method (*Lichtenberger et al.*, 1987), modified for direction finding, is able to produce the necessary frequency resolution. In the following results obtained applying the MFPE method for whistlers are presented; these results confirm the conclusions of Strangeways and Rycroft and also demonstrate the superiority of this special, whistler-based method over the general methods.

4.3.2.3 The MFPE method modified for ground based measurements

The full MFPE method was developed for computing whistler wave normals in satellite measurements. The first detailed analysis of the frequency dependent wave normals of whistlers recorded on a satellite was based on the full MFPE method (see point 4.3.1.4). This method is specific to whistlers, i.e. it uses **a priori** information both in matched filtering and in parameter estimation, and achieves 2~4 times higher accuracy than the general purpose methods, depending on the noise present in with the signal. The computation of the wave normal vector requires at least the three magnetic wave components. In this case a 180° ambiguity remains, which can be resolved using the three electric wave components. Because the main power of the MFPE method is in the determination of the amplitudes and phase constants of a whistler signal of a given frequency rather than the computation of the wave normal vector itself, it is simple to modify this method to the case of a ground based whistler measurement, in which only two horizontal magnetic components and one vertical electric component may be given.

Let us take a right-handed Cartesian frame of reference. Let the x axis be the north, the y axis the west and the z axis the vertical direction. If we have two horizontal magnetic components and the vertical electric component, the complex form of these components is

$$\begin{aligned} B_x &= B_{x0}\,\mathrm{e}^{\mathrm{j}(\omega t+\chi)}, \\ B_y &= B_{y0}\,\mathrm{e}^{\mathrm{j}(\omega t+\xi)}, \\ E_z &= E_{z0}\,\mathrm{e}^{\mathrm{j}(\omega t+\psi)}, \end{aligned} \qquad \textbf{(4.60)}$$

where B_{x0}, B_{y0}, E_{z0} and χ, ξ, ψ are the amplitudes and phase constants for the three components, respectively, which can be computed by the MFPE method.

Following the crossed-loops goniometer approximation (*Bullough and Sagredo,* 1973), the wave is propagated with an azimuthal angle ϕ with respect to the x-axis

$$\mathrm{tg}\,2\phi = \frac{2\hat{B}_y\hat{B}_x\cos\left(\hat{\xi}-\hat{\chi}\right)}{\hat{B}_x^2-\hat{B}_y^2}, \qquad \textbf{(4.61)}$$

where $\hat{B}_x, \hat{B}_y$ and $\hat{\chi}, \hat{\xi}$ are the estimated amplitudes and phases of the two measured magnetic components.

Applying the NPE formalism (*Tsuruda and Hayashi,* 1975) to these components, we obtain the azimuth angle ϕ as

$$\mathrm{tg}\phi = \frac{\hat{B}_y}{\hat{B}_x} \frac{\sin\left(\psi - \hat{\xi}\right)}{\sin\left(\psi - \hat{\chi}\right)}. \tag{4.62}$$

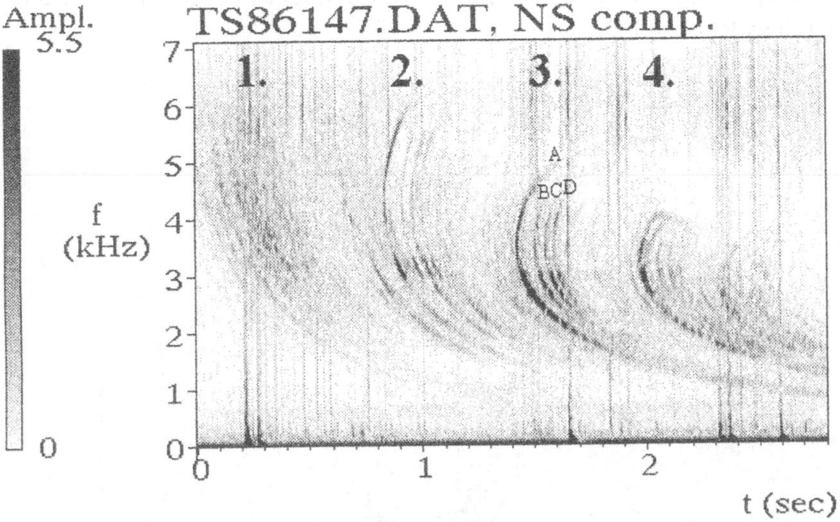

Figure 4.18. Dynamic spectrum of the TS86147 data set, measured at Halley, Antarctica on 10th August 1986, 14:52:09 UT. Courtesy of dr. AJ. Smith, British Antarctic Survey, Cambridge, U.K.

Azimuthal bearing based on two and three measured components:

We analysed a whistler group recorded at Halley, Antarctica (76°S, 27°W, $L \approx 4.3$) on August 10, 1986, beginning at 14:52:09 UT. The dynamic spectrum of this recording is shown in Fig. 4.18 and the analysis of these signals made probable that these whistlers are the results of multipath propagation and were generated probably by a multiple-stroke lightning discharge. If we apply the modified MFPE method that corresponds to the NPE formalism – equation (4.62) –, the azimuthal bearing obtained should only be affected by the (probable) multipath propagation error. However, if the propagating conditions are almost the same, the frequency dependence of azimuthal bearing should also be similar. This is what we expect in computing the azimuth angle for the four components of group 3 in Fig. 4.18. (Remark: In this investigation the theoretical results of oblique propagation – part 2.5 – were not used yet. This is a task for future research, because a possible generation mode of this type of registrations is the oblique propagation of a wide wavefront through inhomogeneous plasmaspheric structures and having nearby exit points, i.e. a localised exit region.)

If we apply the MFPE method modified for two component measurements using (4.61), we expect the combined bearing error to behave like that of the crossed loops goniometer, and, like the three-component version, we expect a very similar frequency dependence for all four traces. Because the four traces are not equally strong, we had to select narrower frequency range for the weaker last three traces B, C and D.

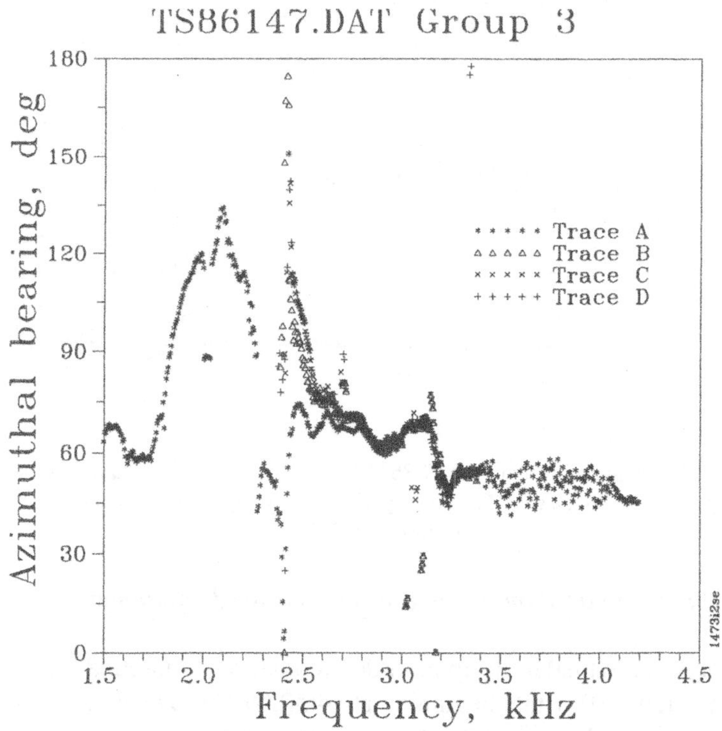

Figure 4.19. Azimuthal bearing of the four traces of group '3' by MFPE method using two horizontal magnetic field components.

The goniometer result is shown in Figure 4.19, it fulfills our expectation: the (ϕ ,f) curves for the four traces are almost identical. Combining this measured azimuth (ϕ $\approx$ 55°) with the L-value of the propagation path, the exit point was found to be located a distance of ~ 140-150 km from the receiving station (see the '+' sign, marked by '7' in Fig. 4.23). The combined systematic bearing error computed by *Strangeways* (1980) shows (Fig. 4.17) that the bearing error between 2 and 3 kHz is almost negligible. The most important factor in the frequency dependence of the bearing error is the propagation distance; the other parameters i.e., ionospheric refractive index, ground conductivity and dielectric constant play less important roles (*Strangeways and Rycroft,* 1980). That is, the averaged value over this frequency range can be taken as the true azimuth angle. (Using the result of *Strangeways* (1980), *Yearby and Smith* (1994) derived the bearing error as a function of frequency and the distance between the exit point and the ground based receiver applying the parameters of Halley Station, British Antarctic Survey. The bearing error was found to be very small between 2 and 3 kHz and up to a distance of 300~400 km, confirming the low error obtained for the bearing in this investigation.)

Figure 4.20 shows the results of the NPE-type computation. Comparing this with Figure 4.21, which shows Strangeways' model calculation on the bearing error of the NPE method, a fairly similar structure can be seen, i.e. the variation of azimuthal bearing is very large, ±180°. If in Figure 4.21 we accept that between 2 and 3 kHz the bearing error is about -10°, and we apply this assumption for Figure 4.20, we obtain a similar value (50-60°) for the azimuth

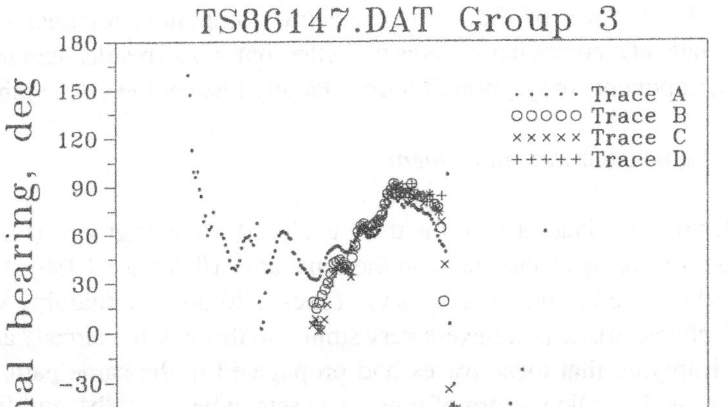

Figure 4.20. Azimuthal bearing of the four traces of group '3' in Fig 4.18 by MFPE method using the two horizontal magnetic and the vertical electric field components.

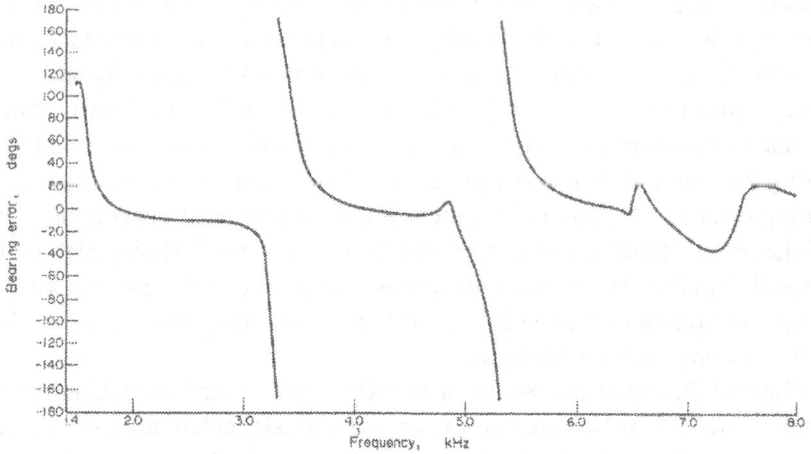

Figure 4.21. Bearing error due to multipath propagation for NPE method for a propagation distance of 150km. Summer night ionosphere and poorly conducting ground. (*Strangeways*, 1980, Fig. 5.)

as we obtained from the two component calculation (Fig. 4.19). Thus, we have concluded that the use of the vertical electric component together with the horizontal magnetic components gives no better, but worse results than using the magnetic components only (though it does largely resolve the 180° ambiguity).

Tracking of a whistler duct movement:

Besides the data set with the three analyzed whistler groups (Fig. 4.18), several other recordings were made on the same day (10 August 1986) at Halley. We selected three other data sets close in time (< 20 min), containing whistler traces with characteristic parameters very similar to those of the already analyzed whistlers, implying that these traces had propagated in the same path. Figure 4.22 shows the dynamic spectra of these data sets, where 'a', 'b', and 'c' mark the chosen traces. The parameters of these four traces obtained by curve fitting are shown in Table 4.1, which also contains the average values obtained for group '3' of the data set TS86147 in Fig. 4.18. Evidently, there is a small systematic variation in the L-value and it increased with time. This suggests that the position of the propagation path, i.e. the "ducting" structure, and thus also the distance between the exit point and the receiver, changed during this period. Hence, the combined bearing error for the goniometric application of the MFPE method had to change also i.e. we expect not only a slightly changing azimuth angle, but that the frequency dependence also should change from trace to trace. Table 4.1 contains the characteristic parameters of the four whistler traces obtained by curve fitting. The quoted errors are standard deviations of the parameters obtained by curve fitting, calculated in a rigorous manner. Of course, systematic errors (biases) in these parameters might also appear, but these will be nearly the same for all whistlers and therefore are not treated or mentioned here in Table 4.1, as being not relevant.

Table 4.1 Characteristic parameters of the four selected whistler traces.

	D_0(s$^{1/2}$)	f_n (Hz)	F_{heq} (Hz)	n_{eq} (cm^{-3})	N_T (×10^{13})cm^{-2}	L
TS86143	89.9±0.4	3540±2	9532±4	222±2.0	3.38±0.033	4.509±0.0010
TS86145	91.4±0.1	3512±1	9456±4	227±0.4	3.49±0.009	4.521±0.0010
TS86146	89.9±0.3	3467±1	9336±2	215±1.0	31.37±0.02	4.540±0.0004
TS86147	92.3±0.1	3407±1	9175±2	220±0.1	3.53±0.003	4.566±0.0001

We have computed the azimuthal bearing for these four traces. Together with the L-values a map can be constructed, showing the movement of the footprint of the whistler exit points (Fig. 4.23). In this figure the times of arrival of the given traces in UT are the following: 14:36:45 UT corresponds to the given trace of data set TS86143, 14:44:56 UT corresponds to the given trace of data set TS86145, 14:48:14 UT corresponds to the given trace of data set TS86146, and 14:52:10 UT corresponds to trace A, group '3' of data set TS86147. The geophysical conclusion is that Figure 4.23 and Table 4.2 show an accelerating cross-L drift; in the equatorial plane it reaches a velocity of 0.4 R_E /hour (corresponding to a magnetospheric eastward electric field of 0.24 mV/m), where R_E is the Earth's radius. A similar magnitude and acceleration in the cross-L drift has been reported by *Carpenter and Seeley* (1976), also based on whistler measurements. Figure 4.23 shows not only cross-*L*, but longitudinal drift as well. The magnitude of the eastward longitudinal drift is ~2 times higher than that of the cross-L drift. This also is in accordance with other measurements based on incoherent radar scattering (*Chappel et al.*, 1970; *Gonzales et al.*, 1980).

Table 4.2 Azimuthal bearing, L-value and the calculated cross-L drift velocity of the four traces

	Time (UT)	ϕ(°)	L	$v_{cross\text{-}L\ drift}$
TS86143	14:36:45	37.6±13.6	4.509±0.0001	
TS86145	14:44:56	43.9±9.4	4.521±0.0001	180 ms^{-1} (0.09 L/h)
TS86146	14:48:14	50.1±5.2	4.540±0.0000	600 ms^{-1} (0.34 L/h)
TS86147	14:52:10	56.1±6.1	4.566±0.0000	700 ms^{-1} (0.40 L/h)

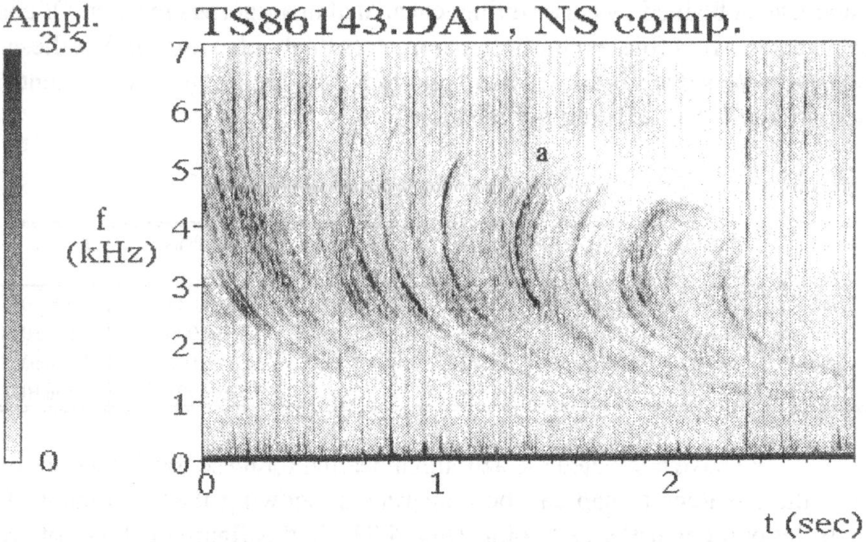

Figure 4.22a. The dynamic spectrum of whistler group recorded at Halley, Antarctica on 10 August 1986 at 14:36:44 UT. The selected trace is 'a'. *

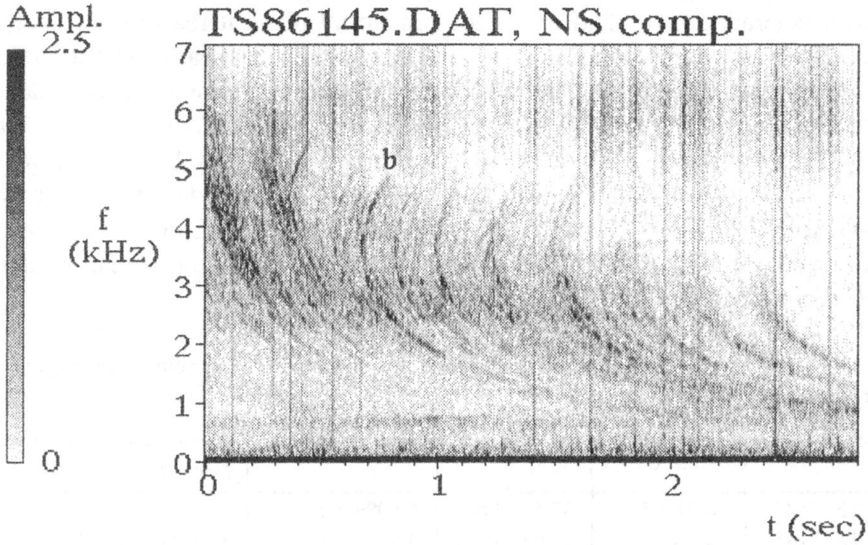

Figure 4.22b. The dynamic spectrum of whistler group recorded at Halley, Antarctica on 10 August 1986 at 14:44:55 UT. The selected trace is 'b'. *

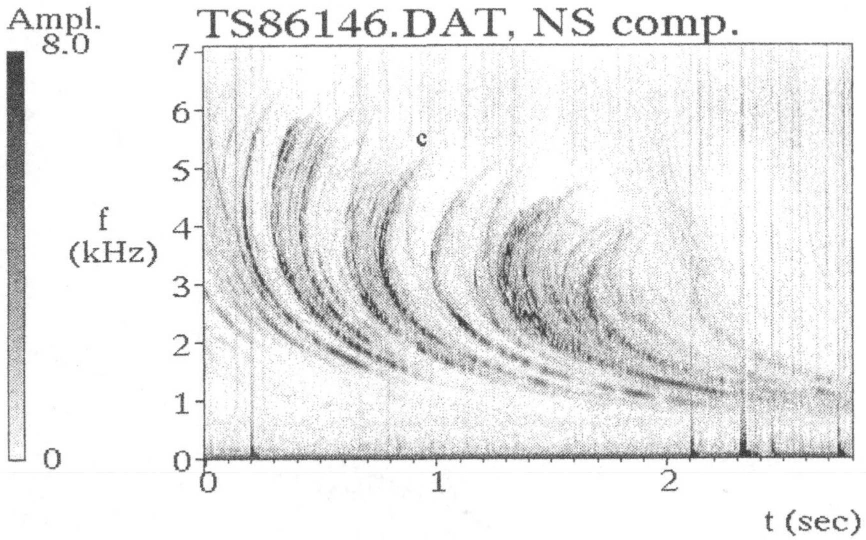

Figure 4.22c. The dynamic spectrum of whistler group recorded at Halley, Antarctica on 10 August 1986 at 14:48:13 UT. The selected trace is 'c'. * (Courtesy of dr. A.J. Smith, British Antarctic Survey, Cambridge, U.K.)

4.4 Benefits of using the real full-wave solutions

In general the benefits produced by using the real full-wave solutions of Maxwell's equations presented in Chapters 1 and 2 are in connection to the existence of the transfer functions of different propagation problems and of other solutions produced by different excitations. It was presented that it is possible to derive the full-wave solutions of different propagation problems of arbitrary shaped signals in a closed form applying the method presented in Chapter 1. After this derivation it is easy to determine the transfer functions of the different propagation situations of the traversed media. Therefore the transfer function opens the way to separate different effects in propagation problems.

4.4.1 Separation of different effects in the received signals

If we know the transfer function of a given propagation problem (i.e. the solution when the excitation is a Dirac-delta distribution), effects in the resultant signal originating from different causes can be separated. During the presentation of this possibility we shall use the whistler-signal propagation in the Earth's magnetosphere as a demonstrative example. The main items in this separation process are the following:

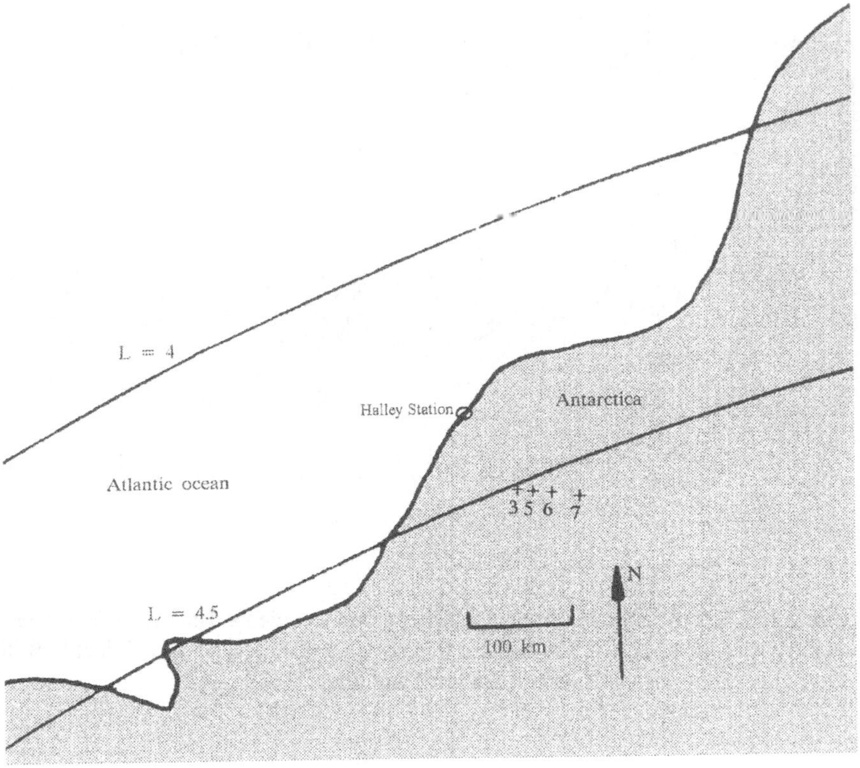

Figure 4.23. A map showing the footprints of the four (selected) whistler traces shown in Figs. 4.18 and 4.22. The numbers seen near the crosses correspond to the time of arrival of the given trace on 10th August, 1986 in UT. (Point 3.: the whistler TS86143, 14:36:44~46 UT, Fig. 4.22a; point 5: the whistler TS86145, 14:44:55~57 UT, Fig. 4.22b; point 6: the whistler TS86146, 14:48:13~15 UT, Fig. 4.22c, point 7: the whistler TS86147, 14:52:09~11 UT, Fig. 4.18.)

– The propagation effects of the traversed medium, inside the validity of the model, are represented by the transfer function response, i.e. $I_{x0}(\omega) = I_0 \triangleq 1$.

– The effects of different (non Dirac-delta) excitation are presented by $I_{x0}(\omega)$ of the given source currents or other excitation spectra.

– Derivation (or separation) of other effects included in the presented (given) model, e.g. the effect of the interference of two whistlers propagating along slightly different paths.

– The detailed comparison of measured and computed signals can produce indications of other (at present unknown) effects, which are not incorporated into the given model.

To reach these goals, it is necessary to apply good methods for signal analysis, e.g. for the investigation of the fine structure of whistlers (see points 4.1 - 4.3). Let us see three examples demonstrating these possibilities.

a) First example: determination of excitation effects

Let us see the characteristics of a whistler measured at Halley Station (Antarctica) on 9 August 1989 (16:17 UT) – this is the No.89221AA whistler (Fig. 4.24). After matched filtering of this measured whistler we can compute whistlers propagating in the same way through the same plasma, i.e. the distance of the magnetic field line from the centre of the Earth, using earth-radius units, in the plane of the equator is $L = 4.246$, the electron density at an altitude of 1000 km is $N = 5780\ \text{cm}^{-3}$, and the plasma will be lossless $(\nu_c = 0)$ or lossy $(\nu_c = 0.2)$ in model calculations using the results presented in points 2.1 and 2.2.

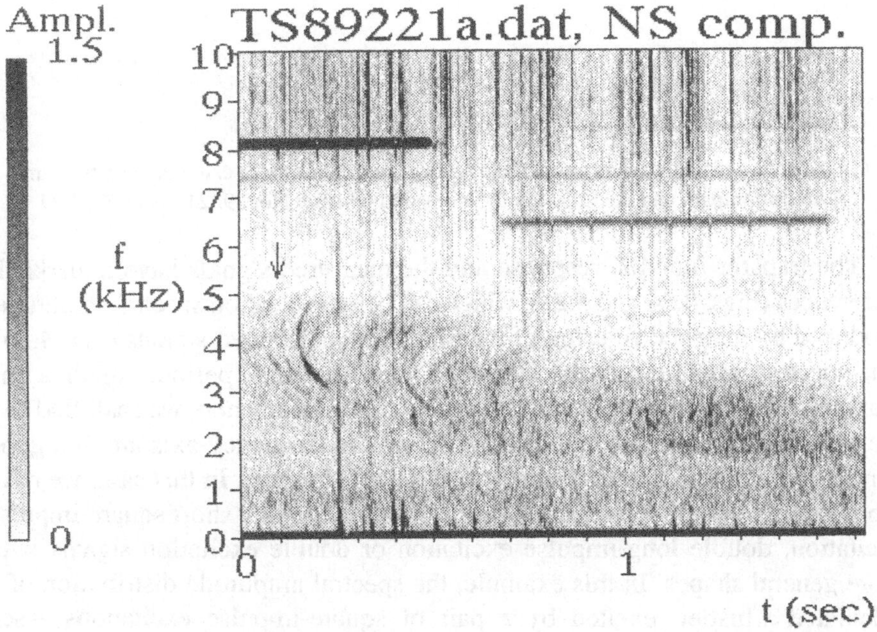

Figure 4.24. The dynamic (FFT) spectrum of the No. 89221AA whistler measured at Halley Station, Antarctica on 9 August, 1989, 16:17 UT. (Courtesy of dr. A.J. Smith, British Antarctic Survey, Cambridge, U.K.). An arrow shows the selected signal.

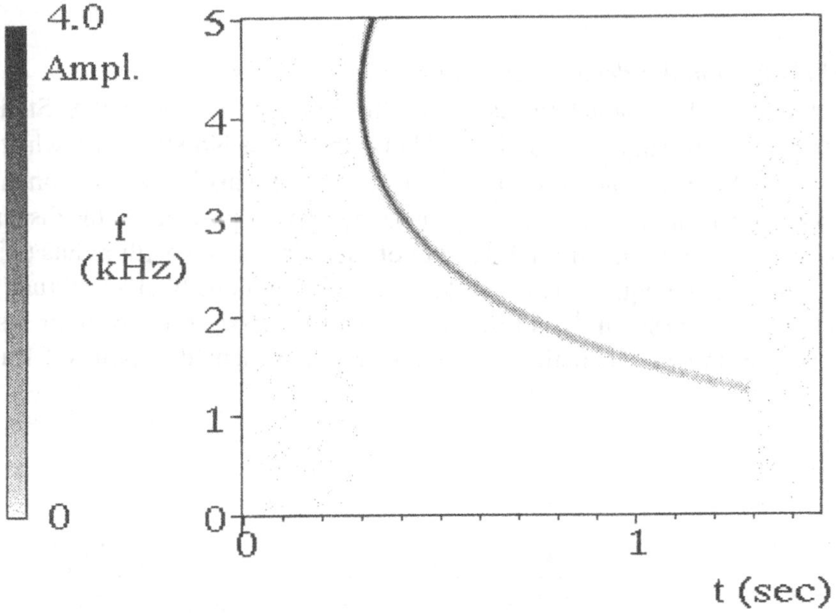

Figure 4.25. Dynamic spectrum of the transfer function response calculated on the same propagation path and condition as the measured whistler No. 89221AA in Fig.4.24

The transfer function response and the measured signals have remarkable differences (Figs. 4.24 and 4.25), which can be seen clear on spectral amplitudes produced by matched filtering of measured and calculated signals (Fig. 4.26). The spectral amplitude of the measured signal presents periodic minima and maxima. Investigating the possible cause of this phenomenon, we shall find that the places of the minima and maxima along the frequency-axis are in a good correlation with the similar picture of "double lightning". In this case, we have more excitation models: double-Dirac excitation, double short square-impulse excitation, double long-impulse excitation or double excitation signals with more general shapes. In this example, the spectral amplitude distribution of a calculated whistler, excited by a pair of square-impulse excitations – see equation (1.74) –, in which the amplitudes of the impulse pair are not equal (the amplitude ratio of the first impulse to the second one is $I_{01} / I_{02} = 2$), is in a good correlation with the measured one. The similarity of the spectral amplitude distribution functions of the calculated and measured signals increases if we suppose that the medium is a lossy one ($\nu_c = 0.2$) – see Fig. 4.27. This means from one side that the No.89221AA whistler (measured at Halley Station, Antarctica) was generated by a non-symmetric double lightning; i.e. it was possible to determine the probable characteristics of the source signal (excitation) of this whistler. On the other hand, the given model and the method gives some valuable information about the parameters of the traversed medium, i.e. the traversed plasma had remarkable loss.

b) Second example: separation of excitation and propagation effects
Let us see three different modelled (calculated) whistlers:

– The first whistler is generated by a *double* Dirac-delta at the plane of $x = x_0$. This excitation is a simple combination of the single Dirac-delta source current density – see equation (1.71a) – , i.e.

$$J_0(x,t) = \left[I_{01}\delta(t) + I_{02}\delta(t-t_0)\right]\delta(x-x_0). \qquad \textbf{(4.63a)}$$

If $I_{01} = I_{02} = 1$, which is the simplest version of this case

$$I_{x0}(\omega) = 2\cos\left(\frac{\omega t_0}{2}\right)e^{-j\omega\frac{t_0}{2}}. \qquad \textbf{(4.63b)}$$

The similarity and difference between this case and the most simple rectangular impulse excitation are remarkable; see equations (1.72) and (4.63). Let the time difference be $t_0 = 2$ ms. The dynamic-spectrum of this computed signal is shown in Fig. 4.28.

– The second whistler is generated by a *simple square-impulse excitation* – see equation (1.72b) – , where the length of this impulse in time is $t_0 = 2.0$ ms. The dynamic spectrum of this computed signal is shown in Fig. 4.29.

– Let the third whistler be a result of the *interference of two transfer functions,* which propagate on very near propagation paths, i.e. $L_1 = 4.654$ and $L_2 = 4.655$, therefore $\Delta L = 0.001$. The dynamic spectrum of this resulting signal is shown in Fig. 4.30.

The qualitative pictures of these signals are similar. Amplitude minima and maxima are present periodically in the FFT-spectrograms and also in the spectral amplitude distribution produced by matched filtering. However, the quantitative parameters of these spectral amplitude distributions – as this can be controlled in Figs. 4.28-4.30 too – are different. The results are seen in Table 4.3.

This means that the differences between these three categories of whistlers are characteristic. There is an exact way to discriminate between phenomena originating from excitations (in this example: double lightning or longer square-impulse type lightning) or from propagation circumstances (in this example: interference of two whistlers propagating along paths with very little differences).

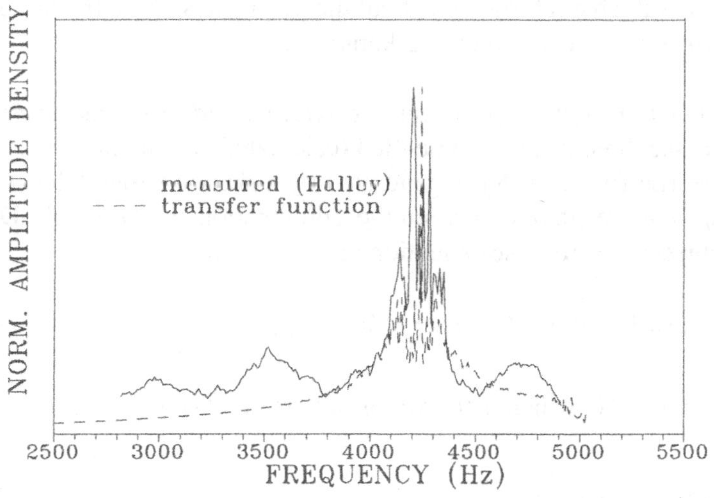

Figure 4.26. Spectral amplitude of the No.89221AA whistler and of the transfer function

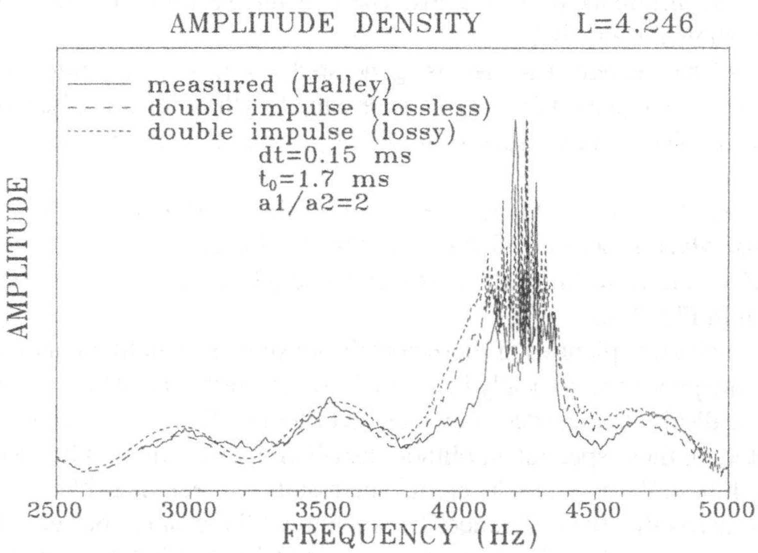

Figure 4.27. Spectral amplitude of the No. 892211AA measured whistler and the computed whistler generated by double-impulse excitation (length of impulses is 0.15 ms; time delay between the two impulses is 1.7 ms; and $I_{01} / I_{02} = 2$) propagating on the same path (L = 4.246; and N_1 =5780) and in first computation the medium is lossless $(\nu_c = 0)$ and in the second one the medium is lossy $(\nu_c = 0.2)$.

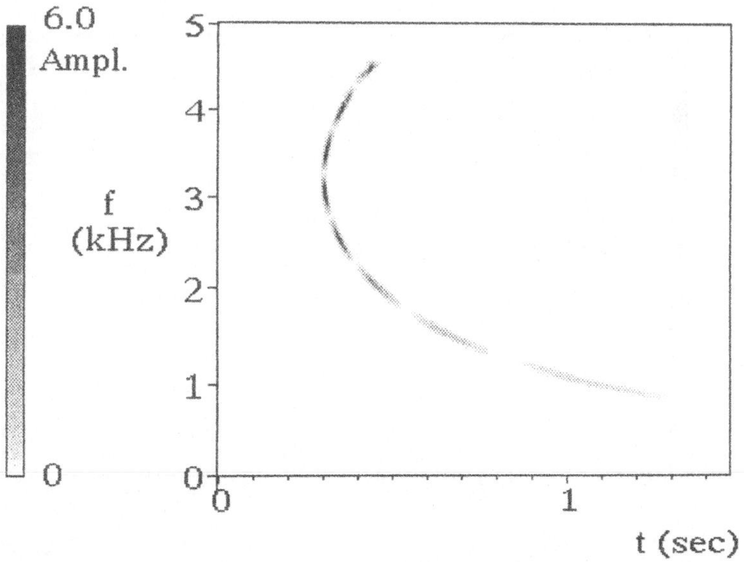

Figure 4.28. The FFT-spectrum of a computed whistler generated by a double-Dirac excitation (L=4.654 and $N_1 = 2550$ cm^{-3}).

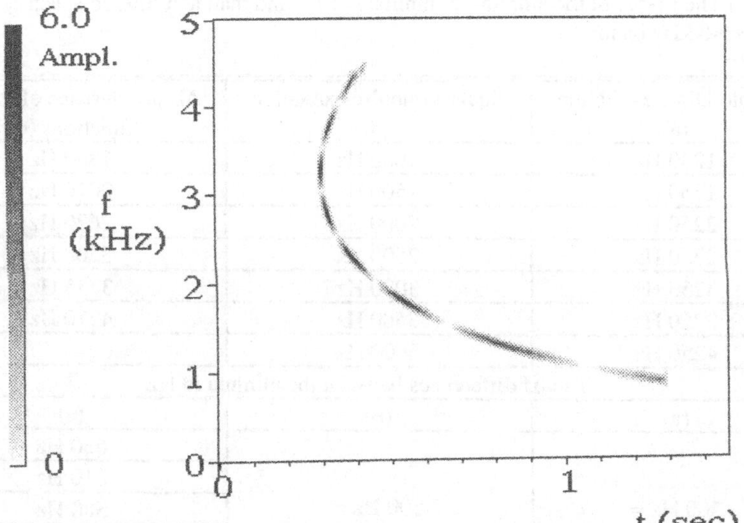

Figure 4.29. The FFT-spectrum of a computed whistler generated by a square-impulse excitation (L=4.654 and $N_1 = 2550$ cm^{-3}).

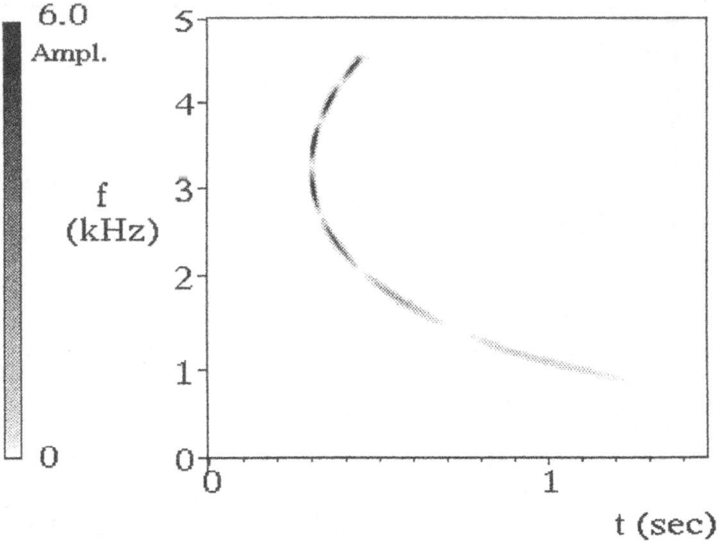

Figure 4.30. The FFT-spectrum of an interference pattern of two transfer functions propagating on very near paths (N_1=2550 cm^{-3}, L_1=4.654; and L_2=4.655)

Table 4.3 The places of the amplitude-minima in FFT and matched filtered spectrograms (in the 1 kHz~5 kHz band)

Double-Dirac excitation (a)	Square-impulse excitation (b)	ΔL interference of transfer functions (c)
1250 Hz	1000 Hz	1380 Hz
1750 Hz	1500 Hz	2010 Hz
2250 Hz	2000 Hz	2620 Hz
2750 Hz	2500 Hz	3200 Hz
3250 Hz	3000 Hz	3755 Hz
3750 Hz	3500 Hz	4210 Hz
4250 Hz	5000 Hz	–
The Δf differences between the minima in Hz		
(a)	(b)	(c)
		630 Hz
		610 Hz
500 Hz =	500 Hz =	580 Hz
		555 Hz
= constant	= constant	455 Hz
		Δf ≠ constant

Parameters:
a) N_1 = 2550 cm^{-3}, L = 4.654, t_0 = 2 ms
b) N_1 = 2550 cm^{-3}, L = 4.654, t_0 = 2 ms
c) N_1 = 2550 cm^{-3}, L_1 = 4.654, L_2 = 4.655

c) Third example: verification of methods for the fine structure analysis of signals

The detailed analysis of the structure of arbitrary shaped signals is very informative as we see it earlier, e.g. in points 4.1~4.3. Let us see for example the matched filtered fine structure of a whistler recorded at Halley Station (Antarctica) on 21 July 1984 – the signal "B" from the TS84010EW whistlers, see in Fig. 4.31 and the matched filtered result in Fig. 4.32. On the matched filtered picture some small sharp splittings are present. The question is the real cause of this type of splittings.

Let us see the matched filtered results of computed signals, e.g. Fig. 4.33 presents the transformed filtered spectrogram of the computed whistler presented in Fig. 4.29. Small sharp splittings appear in this picture too. (Remark: The earlier, generally used, approximate whistler-model applied in the basic version of matched filtering computer code (*Hamar and Tarcsai*, 1991) produces a lower accuracy at the nose of the signal.)

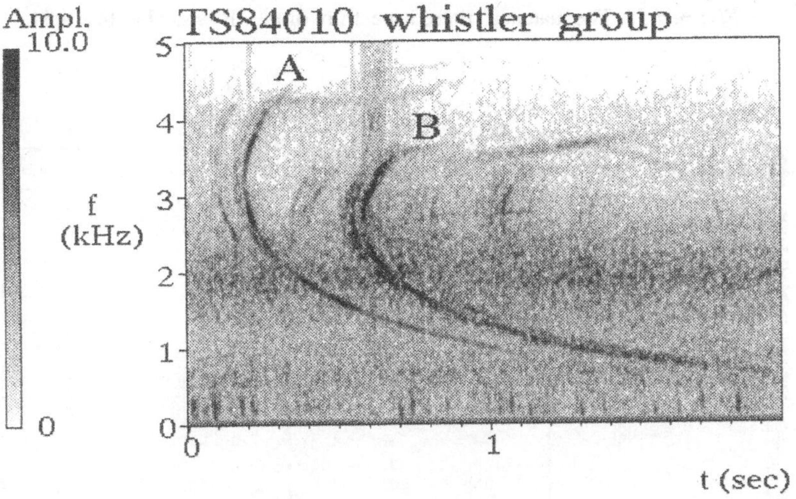

Figure 4.31. The dynamic-spectrum of the TS484010EW whistler group registered at Halley Station (Antarctica) on 21 July 1984, 17:14:13~15UT. (L=4.654; and N_1=2550 cm^{-3}) – Courtesy of dr.. A.J. Smith, British Antarctic Survey, Cambridge, U.K.

Analysing the results, it can be seen that the sharp minima of the original (measured or computed) signal amplitudes appear on the given versions of the matched filtered spectrograms as local splittings of the (whistler) trace. Therefore, we can say that the new solution of Maxwell's equations gives an opportunity to test and verify methods used for fine-structure analysis of signals

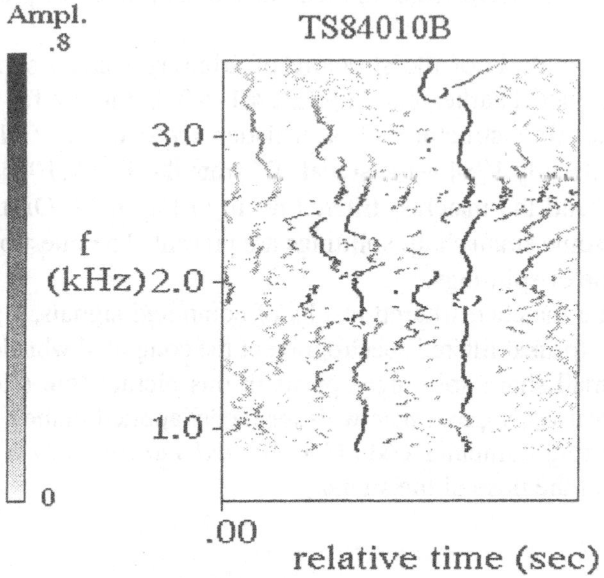

Figure 4.32. The matched filtered spectrum of the "B" signal in Fig. 4.31

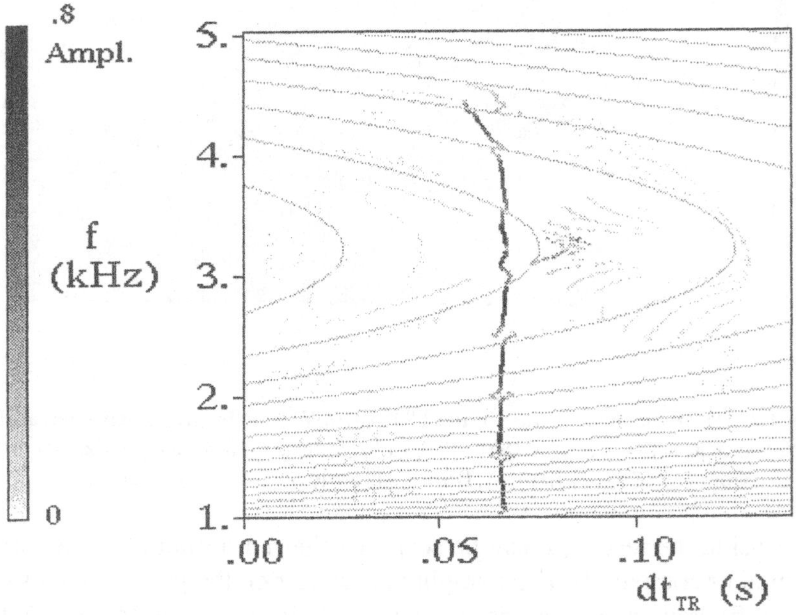

Figure 4.33. The matched filtered spectrum of a calculated whistler generated by a square-impulse excitation (L=4.654 and $N_1 = 2550$ cm^{-3}).

(see more about this in point 4.4.2). Beside this, the application of the new model gives an opportunity for upgrading the analytical methods used in signal analysis.

If we have some idea about the causes of the (fine-) structure of an analysed signal, it is possible to produce the computed version of this signal and – using the same signal processing method – to verify the cause of the interesting event or not.

These examples demonstrate that the new arbitrary shaped solution of Maxwell's equations makes it possible to separate the characteristics of propagation, of excitation and of other effects, to identify "new" (e.g. unknown) phenomena, to test and verify methods of fine structure analysis of signals and to upgrade methods of detailed signal analysis.

4.4.2 Investigation of travel-time residuals

Traditional whistler-research applies the travel-time integrals to characterise the phenomenon, as can be seen in (*Tarcsai*, 1981). However, the real – frequency-depending – time-delays of the measured data show systematic differences (~1ms-10ms) from the values stemming from the travel-time integral equation based monochromatic approximation (*Tarcsai*, 1981). This effect has been supposed to be a very sensitive detector of the model, as the influence of some errors and simplifications in the model may appear in the shape of the travel-time residuals (*Tarcsai*, 1981; *Hamar and Tarcsai*, 1982,

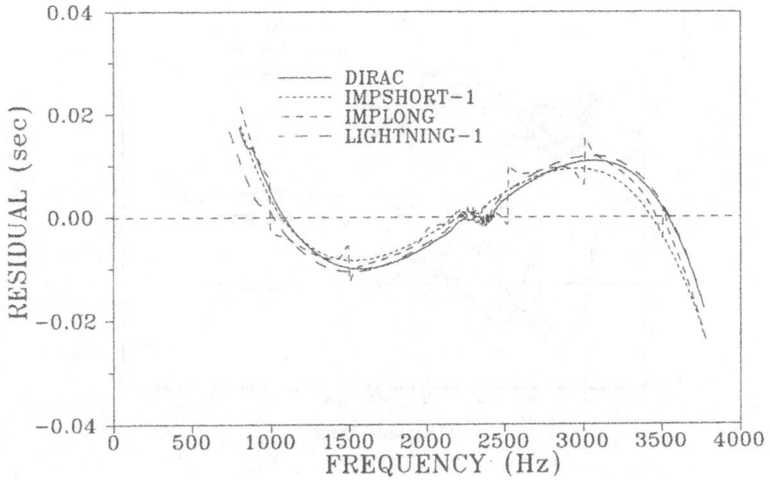

Figure 4.34. Travel time residuals of four calculated whistlers from Figs. 1.7 and 1.8

Strangeways, 1982; *Tarcsai et al.*, 1989; *Hamar et al.*, 1990, 1992; *Lichtenberger et al.*, 1991; *Laird*, 1992; *Sazhin et al.*, 1992) – see Fig. 4.7 or for computed signals Fig. 4.34.

Numerous theories have been based on the detailed fine-structure analysis of the travel-time residuals, regarding assumptions and estimations of different wave-propagation circumstances and plasma-inhomogeneities (fine structure of field-aligned irregularities, "ducts", influence of ion-motions etc.). One of the most important conclusions, originating from the residual-investigations, concerns the general shape of the residual-curves given by averaging many whistler-observations (elimination of the influence of random errors from the measurements). The characteristic form of these curves is a horizontal "S"

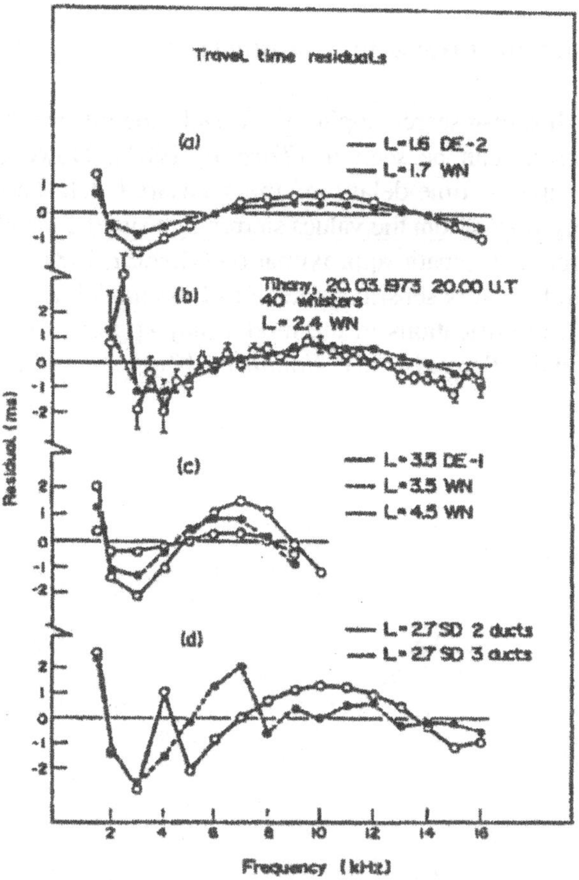

Figure 4.35. Travel time residuals for ducted whistlers, computed by ray tracing for various plasma distributions and at different L-values. in panel (b) a measured curve constructed from 40 whistlers by averaging is also shown (*Tarcsai*, 1981).

along the frequency-axis (see Fig. 4.35). These systematic residuals are assumed to be caused by the negligence of different terms from the applied form of the "refraction index".

Researchers, applying a monochromatic model, executed a detailed analysis of the residuals by comparing the travel-times of model signals (calculated by ray-tracing) from the model based on *Eckersley's* and *Bernard's* formulae. They found that the observed systematic residuals appear mainly as a result of three commonly used approximations. These are the omission of a "+1" term in, and the contribution of ion-motions to, the plasma refractive index, and also the assumption of purely longitudinal propagation rather than the probable monochromatic ray path, which is assumed in the ray-tracing models to execute snake-like excursions back and forth across the ducting structure. In their approach, the greatest part of the residuals is caused by the omission of ionic effects from the refractive index (in peak-to-peak amplitudes the residuals are 4 to 8 times larger for model-whistlers computed with ionic effects than those without that). The "+1" term and the curved ray-path lead to residuals of only several tenths of a ms in their analysis.

Some numerical results and the conclusions:

The application of the new full-wave models in the whistler-research makes it possible to test the signal processing methods and to check the hypotheses (e.g. the more accurate investigations of residuals).

The calculated whistlers are comparable with the measurements and the results of former models, if the same signal processing method is applied, i.e. the matched filtering (see in parts 4.2 and 4.3) and the FIT-method (see in point 4.1.2.1 and *Tarcsai*, 1981). By the aid of these methods the residuals of the full-wave model according to the former one can be determined in the case of Dirac-delta or other excitations.

The comparison of the travel-time residuals of whistlers obtained by the application of the new full-wave model to the ones given by traditional (mainly ray-tracing) model-calculations and to observations has modified the former explanations of the phenomenon.

As is obviously seen in Fig. 4.36, the shape (horizontal “S”) and the order of magnitude (1~4 ms) of the residuals given for strictly longitudinal propagation in the new full-wave model is the same as is known for observations.

In contrast with the former theories, the main reason of this effect is the fundamental inaccuracy of *Bernard's* formula based on monochromatic considerations. Furthermore, the possibility of separation of different effects in the new model has opened the way for detailed investigations of the influence of e.g. the ion-motions and the loss on the travel-time residuals. As is shown in

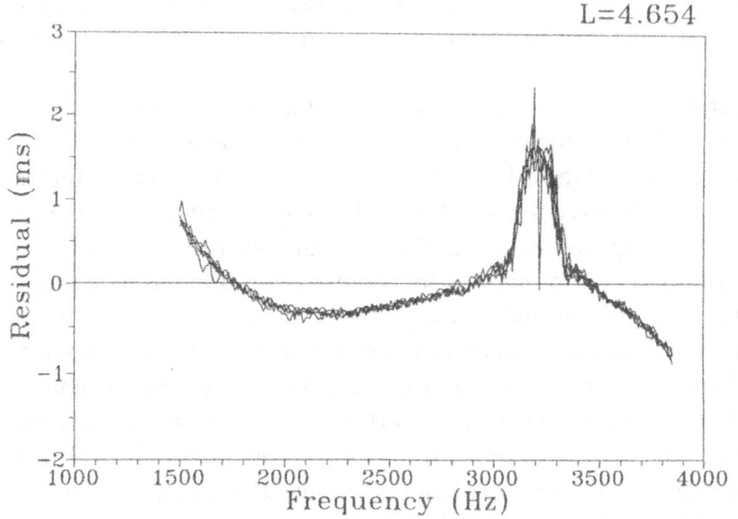

Figure 4.36. Residuals for Dirac-delta excitations — transfer functions — in different plasma-models (spatial inhomogeneity, collision, ion-motions). *: Artefact produced by the present version of the matched filtering process around the nose frequency, caused by the former assumption of the signal-shape (point 4.1.2.2.).

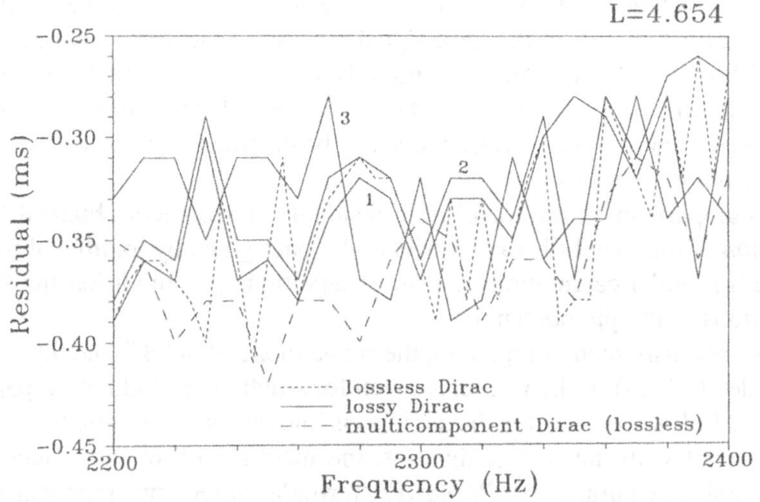

Figure 4.37. Detailed travel time residuals for lossless and lossy plasma cases (1: $v_c = 0.1$, 2: $v_c = 1$, 3: $v_c = 5$ s^{-1}), and for multicomponent, lossless plasma.

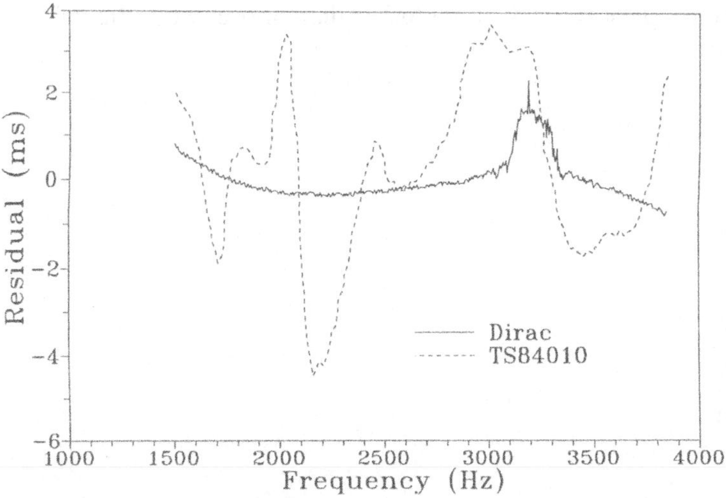

Figure 4.38. Residuals of a measured (Halley, Antarctica, 21 July 1984 – Fig. 4.31) and a calculated whistler (transfer function).

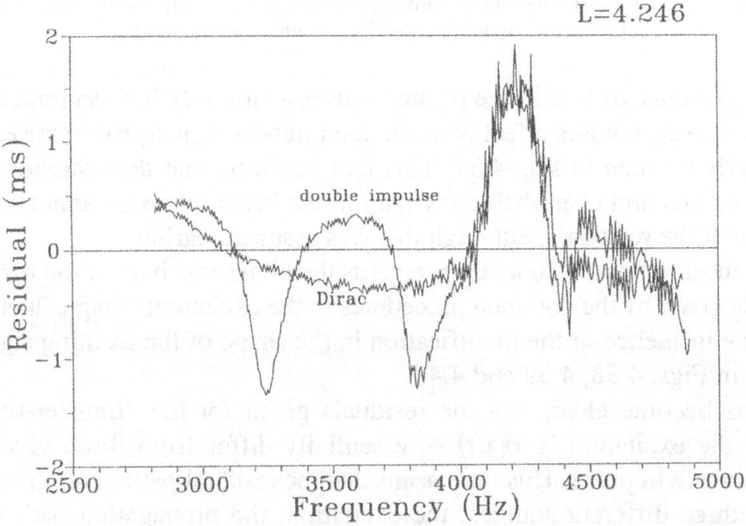

Figure 4.39. Residuals of two calculated whistlers - Dirac-delta and double square-impulse (lossless and lossy, ν_c = 0.2 s^{-1}) excitations.

Fig. 4.37, the ion-motions, taken into consideration by the multi-component plasma (see part 2.4), result in increment of only several (1~2) tenths of a ms, which is much less than was expected on the basis of the former computations (in earlier analysis the residual curves computed with ionic effects gave 4-8 times larger peak-to-peak amplitudes than those with the effects of ions neglected).

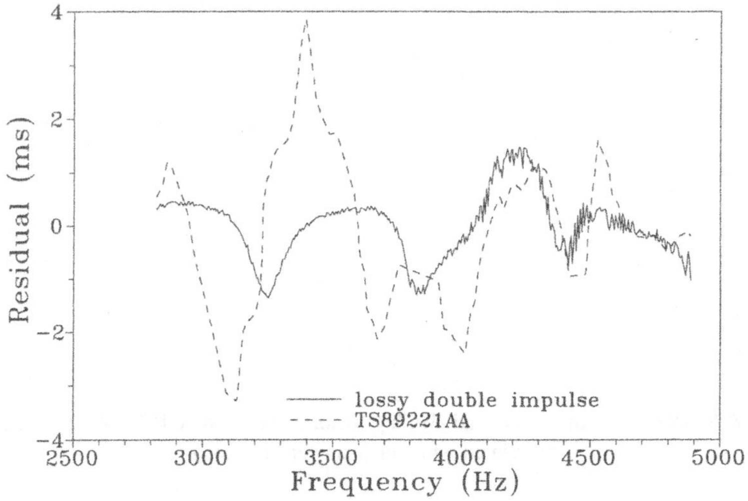

Figure 4.40. Residuals of a measured (Halley, Antarctica, 9 August 1989 – Fig. 4.24) and a calculated whistler (excited by a double square impulse).

The presence of loss in the plasma causes a similarly low decrement in the residual-curves, but this effect is within the limits of signal-processing errors, as can clearly be seen in Fig. 4.37. This fact confirms that the omission of ion-effects, of loss and of probable ray-path in the investigations cannot cause the majority of the residuals, although this was assumed earlier.

The most important conclusion regards the high sensitivity of the travel-time residuals given by the common procedures to the excitation (shape, interference etc.). The influence of the modification in the shape of the exciting signal can be seen in Figs. 4.38, 4.39 and 4.40.

It has become clear, that the residuals given for the "transfer-function" – when the excitation is $\delta(x,t)$ –, essentially differ from those given for a double-square impulse. This fact means that the residual-pattern originates from at least three different sources: the excitation, the propagation path and the measuring (and signal processing) system. Conclusions referring to the propagation circumstances can be deduced only for whistlers that are excited by a Dirac-like signal (transfer function of the medium) or otherwise, the influence of the excitation is highly superimposed on the residual-picture. This result raises the claim to reconsider the former theories and conclusions based on the evaluation of the travel-time residuals.

OUTLOOK

Finally, as an outlook, let the most important applications of the new results and the possibilities of development be mentioned. The flexibility of the model concerning the shape of the excitation, and the fact that the transfer-functions of the investigated plasma-models are known open the way to apply these results in the investigation of VLF-phenomena caused by seismic activity. A very exciting and intensively investigated part of VLF-research is the analysis of non-linear effects in the plasma (e.g. hot plasma models). It seems very important to examine the possibilities of the full-wave method for non-linear plasmas. The problem of the ducted whistler-propagation (duct-theory) will require the application of the full-wave considerations for guided waves (field aligned wave-guides). An open question is the possibility of the generalization of the presented model of obliquely propagating signals for more exact plasma-models containing (strong) spatial inhomogeneity and the influence of the curvature of the inhomogeneous (Earth's) magnetic field.

The flexibility of the model makes it possible to use the presented methods for other (linear) transient wave-propagation problems in higher frequency-domains (TiPP's type and other high frequency phenomena, ionosondes, radar).

As a long-term task, the application of the new results will be very important in the interplanetary VLF- and plasma-investigations. Apart from the exploration of more distant parts of our magnetosphere and the magnetospheres of other planets much experimental work is likely to apply our understanding of the nature of various kinds of waves in order to probe the magnetosphere. A part of this type of work is the use of naturally occurring whistlers to infer the structure of plasmasphere and plasmapause. The study of wave phenomena is a field in which the marriage of theory and observations of natural phenomena has achieved a very high degree of success. It continues to provide a satisfying challenge to both theoretician and experimentalist.

The book presents a summary of a useful and really full-wave solution of Maxwell's equation and some applications of the methods and the results. In Chapter 3 the actuality and importance of these questions is demonstrated. It is clear that the method itself is not a method to find ELF-VLF solutions of wave propagation problems, but it is more general. This method and the solutions are useful without any basic changing in the higher frequency bands too, in which this phenomenological description of the problems are valid and it is not necessary to use the scattering description of the propagation problems.

Therefore, we hope that applying this knowledge will assist the understanding of some Earth related (atmospheric, plasmaspheric, seismic etc.), interplanetary and other planets related phenomena, that the application of it will be successful in some practical applications, i.e. in communication or in buried target detection and in the investigation of global changes too.

* * *

Acknowledgements:

In this book we present some results of actual and long time research work. An essentially important base of this research is a UK-Hungarian cooperation, the long range co-working between the colleagues of the British Antarctic Survey and the Space Research Group of Eötvös University. The authors use a lot of the results of this cooperation and data recorded at the Halley Station of the British Antarctic Survey. The spirit of this successful co-operation is dr. Andy J. Smith our dear friend and we thank Andy's activity and assistance. The authors sincerely thank dr. Andy J. Smith (British Antarctic Survey, Cambridge, U.K.) for providing us with whistler data measured at Halley, Antarctica.

The authors sincerely thank the kind permission of the

– American Geophysical Union (AGU)
for using Figures 2. and 6. of
Storey L.R.O., Lefeuvre F., Parrot M., Cairo L. and Anderson R.R. (1991): Initial survey of the Wave Distribution Functions for plasmaspheric hiss observed by ISEE1; J. Geophys. Res., **96**, 19469.
and for using Figure 3. of
Holden D.N., Manson C.P. and Devenpont J.C. (1995): Satellite observations of transionospheric pulse pairs; Geophys. Research Letters, **22**, 889.

– Elsevier Science
for using Figure 1. of
Watanabe S. and Ondoh P. (1976): Deuteron-whistler and trans-equatorial propagation of the ion-cyclotron whistler; Planet. Space Sci., **24**, 359.
and for using Figures 5., 8., 9. and 11. of
Strangeways H.J. (1980): Systematic errors in VLF direction-finding of whistler ducts, I; J. Atmos. Terr. Phys., **42**, 995.

– Springer-Verlag
for using Figure 10.6 of
Walker A.D.M. (1993): Plasma waves in the magnetosphere; Phys. Chem. Space Planet., **24**, Springer-Verlag, Berlin.

– Stanford University Press

for using Figures 7-1. and 7-34. of
Helliwell R.A. (1965): Whistlers and Related Ionospheric Phenomena; Stanford University Press, Stanford.

The authors express special thanks for the great assistance during the realization of the manuscript to our family members and to the whole staff of the Space Research Group of the Eötvös University (Geophysical Department), especially Péter Bognár, Zoltán Erhardt, Ilona Ferencz-Árkos, Gábor Molnár, Szilárd Pásztor, Péter Steinbach, Ágnes Szabó-Balogh, Balázs Székely, Pál Szemerédy and Gábor Timár. Their work, time and patiency is behind this book.

Appendix A
Determination of integration constants C_y and C_z in equations (1.27)

The second and third component equation in equation (1.26) are

$$\begin{aligned}\frac{\partial v_y}{\partial t}+\omega_b v_z &= \frac{q}{m}E_y,\\ \frac{\partial v_z}{\partial t}-\omega_b v_y &= \frac{q}{m}E_z.\end{aligned} \tag{A.1}$$

Using the results given in (1.27) the new form of (A.1) is:

$$\begin{aligned}&\frac{q}{m}\left\{C_y\frac{\partial}{\partial t}\left[\frac{1}{\omega_b}\int^t\frac{\partial E_y}{\partial\tau}\sin\omega_b(t-\tau)\,\mathrm{d}\tau-\int^t E_z\sin\omega_b(t-\tau)\mathrm{d}\tau\right]+\right.\\ &\left.+C_y\omega_b\frac{\partial}{\partial t}\left[\frac{1}{\omega_b}\int^t\frac{\partial E_z}{\partial\tau}\sin\omega_b(t-\tau)\mathrm{d}\tau+\int^t E_y\sin\omega_b(t-\tau)\mathrm{d}\tau\right]\right\}=\frac{q}{m}E_y,\\ &\frac{q}{m}\left\{C_z\frac{\partial}{\partial t}\left[\frac{1}{\omega_b}\int^t\frac{\partial E_z}{\partial\tau}\sin\omega_b(t-\tau)\,\mathrm{d}\tau+\int^t E_y\sin\omega_b(t-\tau)\,\mathrm{d}\tau\right]-\right.\\ &\left.C_y\omega_b\left[\frac{1}{\omega_b}\int^t\frac{\partial E_y}{\partial\tau}\sin\omega_b(t-\tau)\mathrm{d}\tau-\int^t E_z\sin\omega_b(t-\tau)\mathrm{d}\tau\right]\right\}=\frac{q}{m}E_z.\end{aligned} \tag{A.2}$$

After simplification and derivation by time t we get a relatively simple form of (A.2). This is

$$\begin{aligned}&C_y\int^t\frac{\partial E_y}{\partial\tau}\cos\omega_b(t-\tau)\mathrm{d}\tau-C_y\omega_b\int^t E_z\cos\omega_b(t-\tau)\mathrm{d}\tau+\\ &+C_z\int^t\frac{\partial E_z}{\partial\tau}\sin\omega_b(t-\tau)\mathrm{d}\tau+C_z\omega_b\int^t E_y\sin\omega_b(t-\tau)\mathrm{d}\tau=E_y\ ,\\ &C_z\int^t\frac{\partial E_z}{\partial\tau}\cos\omega_b(t-\tau)\mathrm{d}\tau+C_z\omega_b\int^t E_y\cos\omega_b(t-\tau)\mathrm{d}\tau-\\ &-C_y\int^t\frac{\partial E_y}{\partial\tau}\sin\omega_b(t-\tau)\mathrm{d}\tau-C_y\omega_b\int^t E_z\sin\omega_b(t-\tau)\mathrm{d}\tau=E_z\,.\end{aligned} \tag{A.3}$$

After a simple manipulation of equation (A.3)

$$
\begin{aligned}
&\int^{t}\left[\frac{\partial E_y}{\partial \tau}\cos\omega_b(t-\tau)+\frac{C_z}{C_y}E_y\omega_b\sin(t-\tau)\right]\mathrm{d}\tau+\\
&+\int^{t}\left[\frac{C_z}{C_y}\frac{\partial E_z}{\partial \tau}\sin\omega_b(t-\tau)-E_z\omega_b\cos(t-\tau)\right]\mathrm{d}\tau=\frac{E_y}{C_y},\\
&\int^{t}\left[\frac{\partial E_z}{\partial \tau}\cos\omega_b(t-\tau)+\frac{C_y}{C_z}E_z\omega_b\sin(t-\tau)\right]\mathrm{d}\tau-\\
&-\int^{t}\left[\frac{C_y}{C_z}\frac{\partial E_y}{\partial \tau}\sin\omega_b(t-\tau)-E_y\omega_b\cos(t-\tau)\right]\mathrm{d}\tau=\frac{E_z}{C_z}.
\end{aligned}
\qquad \textbf{(A.4)}
$$

Multiplying the left side of the first equation in (A.4) by E_z/C_z and the left side of the second equation in (A.4) by E_y/C_y it can be seen that the new equation can be fulfilled only if

$$
C_y = C_z \triangleq C \;. \qquad \textbf{(A.5)}
$$

Applying (A.5) in (A.4) the new relations are well known, i.e.

$$
\begin{aligned}
&\int^{t}\left[\frac{\partial E_y}{\partial \tau}\cos\omega_b(t-\tau)+E_y\omega_b\sin\omega_b(t-\tau)\right]\mathrm{d}\tau+\\
&+\int^{t}\left[\frac{\partial E_z}{\partial \tau}\sin\omega_b(t-\tau)-E_z\omega_b\cos\omega_b(t-\tau)\right]\mathrm{d}\tau=\frac{E_y}{C},\\
&\int^{t}\left[\frac{\partial E_z}{\partial \tau}\cos\omega_b(t-\tau)+E_z\omega_b\sin\omega_b(t-\tau)\right]\mathrm{d}\tau-\\
&-\int^{t}\left[\frac{\partial E_y}{\partial \tau}\sin\omega_b(t-\tau)-E_y\omega_b\cos\omega_b(t-\tau)\right]\mathrm{d}\tau=\frac{E_z}{C}.
\end{aligned}
\qquad \textbf{(A.6)}
$$

From equations (A.6)

$$
\begin{aligned}
&\int^{t}\frac{\partial}{\partial \tau}[E_y\cos\omega_b(t-\tau)]\mathrm{d}\tau+\int^{t}\frac{\partial}{\partial \tau}[E_z\sin\omega_b(t-\tau)]\mathrm{d}\tau=\frac{E_y}{C},\\
&\int^{t}\frac{\partial}{\partial \tau}[E_z\cos\omega_b(t-\tau)]\mathrm{d}\tau-\int^{t}\frac{\partial}{\partial \tau}[E_y\sin\omega_b(t-\tau)]\mathrm{d}\tau=\frac{E_z}{C}.
\end{aligned}
\qquad \textbf{(A.7)}
$$

After integration on the left side of equations (A.7)

$$E_y(t)\cos(\omega_b\cdot 0)+E_z(t)\sin(\omega_b\cdot 0)=\frac{1}{C}E_y(t),$$
$$E_z(t)\cos(\omega_b\cdot 0)+E_y(t)\sin(\omega_b\cdot 0)=\frac{1}{C}E_z(t),$$

from which

$$E_y(t)=\frac{1}{C}E_y(t),\qquad E_z(t)=\frac{1}{C}E_z(t), \tag{A.8}$$

and therefore

$$C=1. \tag{A.9}$$

Appendix B
Derivation of the monochromatic case from the general equations

A special case of the general solution is the version in which *the form of solution is an endless monochromatic signal,* i.e.

$$\overline{E}=\overline{E}_0 \mathrm{e}^{\mathrm{j}\omega t}, \qquad \text{where} \quad \omega=\text{constant}. \tag{B.1}$$

In this case, the form of the first Maxwell's equation is

$$\overline{\nabla}\times\overline{H}=\overline{J}+\varepsilon_0\frac{\partial\overline{E}}{\partial t}=\overline{J}+\mathrm{j}\omega\varepsilon_0\overline{E}. \tag{B.2}$$

Let us define the electric displacement as

$$\mathrm{j}\omega\overline{D}\triangleq\overline{J}+\mathrm{j}\omega\varepsilon_0\overline{E}. \tag{B.3}$$

The form of the current density $\overline{J}$ in this spectral case could derive using equations (1.28) and (B.1) together. The first current density component is

$$J_x=\varepsilon_0\omega_p^2\int^t E_x \mathrm{d}\tau=\varepsilon_0\omega_p^2\int^t E_{x0}\mathrm{e}^{\mathrm{j}\omega\tau}\mathrm{d}\tau=-\mathrm{j}\omega\varepsilon_0\frac{\omega_p^2}{\omega^2}. \tag{B.4}$$

Let us use, in the following, the notations

$$\alpha^2=\frac{\omega_p^2}{\omega^2} \qquad \text{and} \qquad \beta=\frac{\omega_b}{\omega}. \tag{B.5}$$

Therefore

$$J_x=\mathrm{j}\omega\varepsilon_0(-\alpha^2)E_x. \tag{B.6}$$

The second component of the current density is

$$
\begin{aligned}
J_y &= \varepsilon_0\omega_p^2\left\{\frac{1}{\omega_b}\int^t \frac{\partial\left(E_{z0}e^{j\omega\tau}\right)}{\partial\tau}\sin\omega_b(t-\tau)\cdot d\tau - \int^t \left(E_{z0}e^{j\omega\tau}\right)\sin\omega_b(t-\tau)\cdot d\tau\right\} = \\
&= \varepsilon_0\omega_p^2\left\{\frac{j\omega}{\omega_b}E_{y0}\left[\sin\omega_b t\int^t e^{j\omega\tau}\cos\omega_b\varepsilon\cdot d\tau - \cos\omega_b t\int^t e^{j\omega\tau}\sin\omega_b\tau\cdot d\tau\right] - \right. \\
&\left. - E_{z0}\left[\sin\omega_b t\int^t e^{j\omega\tau}\cos\omega_b\tau\cdot d\tau - \cos\omega_b t\int^t e^{j\omega\tau}\sin\omega_b\tau\cdot d\tau\right]\right\}.
\end{aligned}
\tag{B.7}
$$

It is known that

$$
\begin{aligned}
\int^x e^{ax}\cos bx\cdot dx &= \frac{e^{ax}}{a^2+b^2}(a\cos bx + b\sin bx) + C, \\
\int^x e^{ax}\sin bx\cdot dx &= \frac{e^{ax}}{a^2+b^2}(a\sin bx + b\cos bx) + C.
\end{aligned}
\tag{B.8}
$$

Applying (B.8) in (B.7) and simplifying the result

$$
\begin{aligned}
J_y &= \varepsilon_0 E_{y0}e^{j\omega t}j\omega\frac{\omega_p^2}{\omega_b^2-\omega^2} - \varepsilon_0 E_{z0}e^{j\omega t}\frac{\omega_b\omega_p^2}{\omega_b^2-\omega^2} = \\
&= \varepsilon_0\left[-\frac{\alpha^2}{1-\beta^2}j\omega E_y - j\beta\frac{\alpha^2}{1-\beta^2}j\omega E_z\right].
\end{aligned}
\tag{B.9}
$$

In the same manner

$$
J_z = \varepsilon_0\left[-\frac{\alpha^2}{1-\beta^2}j\omega E_z + j\beta\frac{\alpha^2}{1-\beta^2}j\omega E_y\right].
\tag{B.10}
$$

Applying equations (B.6), (B.9) and (B.10) in equation (B.3) we get back the well known solution of monochromatic case (e.g. *Allis et al.,* 1963), i.e.

$$
\begin{aligned}
D_x &= j\omega\varepsilon_0(1-\alpha^2)E_x \stackrel{\Delta}{=} j\omega\varepsilon_0\varepsilon_{II}E_x, \\
D_y &= j\omega\varepsilon_0\left[\left(1-\frac{\alpha^2}{1-\beta^2}\right)E_y - j\left(1-\frac{\alpha^2}{1-\beta^2}\right)E_z\right] \stackrel{\Delta}{=} j\omega\varepsilon_0\left[\varepsilon_\perp E_y - j\varepsilon_x E_z\right], \\
D_z &= j\omega\varepsilon_0\left[\varepsilon_\perp E_z + j\varepsilon_x E_y\right].
\end{aligned}
\tag{B.11}
$$

The final and well known form of (B.11) is

$$\overline{D} = \mathrm{j}\omega\overline{\overline{\varepsilon}}(\omega)\overline{E}\,, \tag{B.12a}$$

where the permittivity is

$$\overline{\overline{\varepsilon}}(\omega) = \varepsilon_0 \begin{bmatrix} \varepsilon_{II} & 0 & 0 \\ 0 & \varepsilon_{\perp} & -\mathrm{j}\varepsilon_x \\ 0 & +\mathrm{j}\varepsilon_x & \varepsilon_{\perp} \end{bmatrix} \tag{B.12b}$$

in the given coordinate-system.

Appendix C
Fulfilment of the conservation of wave crests

The principle of the conservation of wave crests is well known in the case of signals

$$\overline{G} = \overline{G}_0 \mathrm{e}^{\mathrm{j}\phi}, \tag{C.1}$$

where $\overline{G}_0$ is a real number (value). Let us use the following definitions

$$\overline{k} = -\overline{\nabla}\phi \qquad \text{and} \qquad \omega = \frac{\partial \phi}{\partial t}. \tag{C.2}$$

In this case the equations

$$\begin{aligned} &\overline{\nabla} \times \overline{k} = 0, \\ &\frac{\partial \overline{k}}{\partial t} + \overline{\nabla}\omega = 0 \end{aligned} \tag{C.3}$$

are valid. Equations (C.3) are the conservation law of wave crests, which means that the frequency of the given signal can change but only in such a way which is defined by (C.3).

If the form of a generalised phase is

$$\begin{aligned} \phi = &\int^{t} \Omega(x,y,z,\tau)\mathrm{d}\tau - \int^{x} K_x(\xi,y,z,t)\mathrm{d}\xi - \\ &- \int^{y} K_y(x,\eta,z,t)\mathrm{d}\eta - \int^{z} K_z(x,y,\zeta,t)\mathrm{d}\zeta, \end{aligned} \tag{C.4}$$

the equations (C.3) remain valid. Using (C.2)

$$
\begin{aligned}
\omega &= \Omega - \int^{x} \frac{\partial K_x}{\partial t} \mathrm{d}\xi - \int^{y} \frac{\partial K_y}{\partial t} \mathrm{d}\eta - \int^{z} \frac{\partial K_z}{\partial t} \mathrm{d}\zeta , \\
k_x &= -\int^{t} \frac{\partial \Omega}{\partial x} \mathrm{d}\tau + K_x + \int^{y} \frac{\partial K_y}{\partial x} \mathrm{d}\eta + \int^{z} \frac{\partial K_z}{\partial x} \mathrm{d}\zeta , \\
k_y &= -\int^{t} \frac{\partial \Omega}{\partial y} \mathrm{d}\tau + \int^{x} \frac{\partial K_x}{\partial y} \mathrm{d}\xi + K_y + \int^{z} \frac{\partial K_z}{\partial y} \mathrm{d}\zeta , \\
k_z &= -\int^{t} \frac{\partial \Omega}{\partial z} \mathrm{d}\tau + \int^{x} \frac{\partial K_x}{\partial z} \mathrm{d}\xi + \int^{y} \frac{\partial K_y}{\partial z} \mathrm{d}\eta + K_z .
\end{aligned}
\qquad \text{(C.5)}
$$

By applying equations (C.5) in (C.3) it is evident that the equations (C.3) remain valid. Let us see the test of this in one-one component equations, i.e.

$$
\overline{\nabla} \times \overline{k} = 0 \quad \Rightarrow \quad \left(\frac{\partial k_z}{\partial y} - \frac{\partial k_y}{\partial z} \right) \overline{e}_1 , \qquad \text{etc.}
$$

$$
\frac{\partial \overline{k}}{\partial t} + \overline{\nabla} \omega = 0 \quad \Rightarrow \quad \left(\frac{\partial k_x}{\partial t} - \frac{\partial \omega}{\partial x} \right) \overline{e}_1 , \qquad \text{etc.}
$$

Therefore

$$
\begin{aligned}
\frac{\partial k_z}{\partial y} - \frac{\partial k_y}{\partial z} = &-\int^{t} \frac{\partial^2 \Omega}{\partial y \partial z} \mathrm{d}\tau + \int^{x} \frac{\partial^2 K_x}{\partial y \partial z} \mathrm{d}\xi + \frac{\partial K_y}{\partial z} + \frac{\partial K_z}{\partial y} + \\
&+ \int^{t} \frac{\partial^2 \Omega}{\partial z \partial y} \mathrm{d}\tau - \int^{t} \frac{\partial^2 K_x}{\partial z \partial y} \mathrm{d}\xi - \frac{\partial K_y}{\partial z} - \frac{\partial K_z}{\partial y} = 0
\end{aligned}
$$

and

$$
\begin{aligned}
\frac{\partial k_x}{\partial t} + \frac{\partial \omega}{\partial x} = &-\frac{\partial \Omega}{\partial x} + \frac{\partial K_x}{\partial t} + \int^{y} \frac{\partial^2 K_y}{\partial t \partial x} \mathrm{d}\eta + \int^{z} \frac{\partial^2 K_z}{\partial t \partial x} \mathrm{d}\zeta + \\
&+ \frac{\partial \Omega}{\partial x} + \frac{\partial K_x}{\partial t} - \int^{y} \frac{\partial^2 K_y}{\partial x \partial t} \mathrm{d}\eta - \int^{z} \frac{\partial^2 K_z}{\partial x \partial t} \mathrm{d}\zeta = 0.
\end{aligned}
\qquad \text{(C.6)}
$$

The situation is identical in the case of the other component equations.

Appendix D
Inverse Laplace transform of the excited signals in half-space "1"

As is known

$$E_{1z}(p,s)=\frac{Z_0}{2}\frac{J_0(p,s)}{p-\frac{s}{c}}-\frac{Z_0}{2}\frac{J_0(p,s)}{p+\frac{s}{c}}. \tag{1.50}$$

From equation (1.50) the inverse-transformational task is

$$F(p,s)=\mp\frac{A_0(p,s)}{p\pm\frac{s}{c}} \qquad \Rightarrow \qquad f(x,t). \tag{D.1}$$

The roots are

$$p_{1,2}=\mp\frac{s}{c}, \qquad \text{i.e.} \qquad k_{1,2}=\pm\frac{\omega}{c}=\pm k_0. \tag{D.2}$$

From another point of view

$$\frac{A_0(p,s)}{p\pm\frac{s}{c}}=\mp A_0(p,s)\cdot g(p,s) \tag{D.3}$$

and therefore

$$f(x,t)=\mp\int_0^\infty\int_0^\infty A_0(\xi,\tau)\cdot g(x-\xi,t-\tau)\cdot \mathrm{d}\xi\cdot\mathrm{d}\tau. \tag{D.4}$$

In our case the signal does not exist in the half-spaces $x>x_0$ and $t>t_0$, where $t_0<\infty$. Therefore

$$f(x,t)=\mp\int_0^{x_0}\int_0^{t_0} A_0(\xi,\tau)\cdot g(x-\xi,t-\tau)\cdot \mathrm{d}\xi\cdot\mathrm{d}\tau. \tag{D.5}$$

If we see the functional $g(p,s)$ than

$$g(p,s)=\frac{1}{p\pm\frac{s}{c}} \qquad \text{and therefore}$$

$$g(x,s)=1\cdot \mathrm{e}^{\mp\frac{s}{c}x}.$$

From this

$$g(x,t)=\delta\left(t\mp\frac{x}{c}\right), \tag{D.6}$$

where δ is the Dirac-distribution.

Applying (D.6) in equation (D.5) we get the final form of the solution. The signal propagating in the $+x$ direction is

$$g(x,t)=\delta\left(t-\frac{x}{c}\right).$$

Therefore

$$E_{1_z}(x,t)=-\frac{Z_0}{2}\int_0^{x_0}\int_0^{t_0} J_0(\xi,\tau)\cdot\delta\left[\left(t-\frac{x}{c}\right)-\left(\tau-\frac{\xi}{c}\right)\right]\cdot \mathrm{d}\xi\cdot \mathrm{d}\tau\,.$$

In this relation $\delta(0)$ defines the whole value of the integral. The argument of δ will be zero if

$$t-\tau-\frac{x}{c}+\frac{\xi}{c}=0$$

and

$$\tau=t-\frac{x-\xi}{c}. \tag{D.7}$$

Therefore

$$E_{1_z}(x,t)=-\frac{Z_0}{2}\int_0^{x_0} J_0\left(\xi,t-\frac{x-\xi}{c}\right)\cdot \mathrm{d}\xi. \tag{D.8}$$

Appendix E
Summary of the Wentzel-Kramer-Brillouin (W.K.B.) method

A sufficiently accurate solution of Maxwell's equations can be obtained in the case of weakly inhomogeneous (quasi-homogeneous) media by the application of the W.K.B. method (e.g. in *Budden*, 1966; *Ferencz*, 1996). This approximation is also useful in the case of anisotropic media, if the direction of propagation is known and strictly determined. (A solution obtained for a strongly inhomogeneous, anisotropic medium gives back the W.K.B.-solution in the first approximation for quasi-homogeneous case, assuming a definitive propagation-direction parallel with the gradient of the inhomogeneity).

The basic principle of the method can be demonstrated by assuming a weakly inhomogeneous, isotropic medium, and the solution looked for has the well-known, monochromatic form

$$\overline{G} = \overline{G}_0 \cdot \mathrm{e}^{\mathrm{j}(\omega t - \varphi)} . \tag{E.1}$$

A common approximation is in a quasi-homogeneous medium

$$\overline{\overline{\varepsilon}}_{\mathrm{INH}}(\overline{r}) \rightarrow \varepsilon_{\mathrm{H}}(\overline{r}), \tag{E.2}$$

and

$$\overline{D} = \varepsilon_{\mathrm{H}}(\overline{r}) \cdot \overline{E} , \tag{E.3}$$

where $\overline{\overline{\varepsilon}}_{\mathrm{INH}}$ is the permittivity tensor of an inhomogeneous medium and ε_{H} is the permittivity of an isotropic homogeneous medium.

Assuming that in the medium to be examined

$$\overline{B} = \mu_0 \overline{H} \tag{E.4}$$

further, that the gradient of inhomogeneity is parallel with the z-axis of the coordinate-system

$$\overline{D} = \varepsilon_{\mathrm{H}}(z) \overline{E} . \tag{E.5}$$

The solution of Maxwell's equations is supposed to be a plane-wave, so in a medium specified above, the only possible signal must propagate to z-direction.

Let the common definition be introduced for refraction index

$$n^2(z) = \frac{\varepsilon_{\mathrm{H}}(z)}{\varepsilon_0}. \tag{E.6}$$

Let the inhomogeneous (generalised) propagation vector be introduced for the phase-function φ of the solution

$$\overline{K} = \overline{\nabla}\varphi(z) = K\overline{e}_z, \tag{E.7}$$

$$\varphi(z) = \int^{z} K(z)\mathrm{d}z. \tag{E.8}$$

Substituting the fields in (E.1) into Maxwell's equations

$$\begin{aligned} \overline{\nabla}\times\overline{H} &= \mathrm{j}\omega\,\varepsilon_{\mathrm{H}}(z)\overline{E}, \\ \overline{\nabla}\times\overline{E} &= \mathrm{j}\omega\,\mu_0\overline{H} \end{aligned} \tag{E.9}$$

are given. Rearranging (E.9), $E_z = H_z = 0$ can be obtained on the common way, and

$$\begin{aligned} &\frac{\partial H_y}{\partial z} = -\mathrm{j}\omega\,\varepsilon_{\mathrm{H}}(z)E_x, && \frac{\partial H_x}{\partial z} = \mathrm{j}\omega\,\varepsilon_{\mathrm{H}}(z)E_y, \\ &&\text{and}\\ &\frac{\partial E_x}{\partial z} = -\mathrm{j}\omega\,\mu_0 H_y, && \frac{\partial E_y}{\partial z} = \mathrm{j}\omega\,\mu_0 H_x. \end{aligned} \tag{E.10}$$

These two independent equation-systems deliver the orthogonal EM and HM solution-couples. Assuming rectangular incidence, the two solutions are identical, so it is sufficient to investigate only one of them. It derives from (E.10) in the common way, that

$$\frac{\partial^2 E_y}{\partial z^2} + \omega^2\varepsilon_{\mathrm{H}}(z)\mu_0 E_y = 0. \tag{E.11}$$

An approximate solution of (E.11) is

$$K = \pm\omega\sqrt{\varepsilon_{\mathrm{H}}(z)\mu_0} = \pm k_0 n(z). \qquad \textbf{(E.12)}$$

Therefore

$$\begin{aligned} E_y &= E_0 \cdot \mathrm{e}^{\mathrm{j}\left[\omega t \mp k_0 \int^z n(z)\mathrm{d}z\right]}, \\ H_x &= \mp\frac{E_0}{Z_0} n(z)\cdot \mathrm{e}^{\mathrm{j}\left[\omega t \mp k_0 \int^z n(z)\mathrm{d}z\right]}. \end{aligned} \qquad \textbf{(E.13)}$$

The complex Poynting-vector of the signal is

$$\bar{S}_K = \frac{1}{2}\bar{E}\times\bar{H}^*,$$

where

$$S_K \sim n(z)\cdot E_0^2. \qquad \textbf{(E.14)}$$

In a first approximation let the power-density of the signal be considered to be constant, neglecting the part of the energy reflected and scattered on the inhomogeneity. So $S_K \cong$ constant and

$$E_0 \sim \frac{\text{constant}}{\sqrt{n(z)}}, \text{ where } n(z) \neq 0 \qquad \textbf{(E.15)}$$

is yielded. This is the essence of the W.K.B. philosophy

Writing back (E.15) to (E.13) the so-called W.K.B. solution-couple propagating to $\pm z$ direction can be obtained

$$\begin{aligned} E_{y1} &= \frac{E_{01}}{\sqrt{n(z)}}\cdot \mathrm{e}^{\mathrm{j}\left[\omega t - k_0 \int^z n(z)\mathrm{d}z\right]}, & E_{y2} &= \frac{E_{02}}{\sqrt{n(z)}}\cdot \mathrm{e}^{\mathrm{j}\left[\omega t + k_0 \int^z n(z)\mathrm{d}z\right]}, \\ H_{x1} &= -\frac{E_{01}}{Z_0}\sqrt{n(z)}\cdot \mathrm{e}^{\mathrm{j}\left[\omega t - k_0 \int^z n(z)\mathrm{d}z\right]}, & H_{x2} &= \frac{E_{02}}{Z_0}\sqrt{n(z)}\cdot \mathrm{e}^{\mathrm{j}\left[\omega t + k_0 \int^z n(z)\mathrm{d}z\right]}. \end{aligned} \qquad \textbf{(E.16)}$$

The coupled waves

In the solving process outlined above the new effects appearing in presence of inhomogeneity (reflection and scattering of the signal) are not considered. To precise the result, the fact must be taken into account, that the two signals propagating over against each other (forward and backward) are present simultaneously in the medium, they are not independent of each other, and they exist only together.

Let the signal be looked for in the following form

$$
\begin{aligned}
E_{y1,2} &= \frac{A_{01,2}(z)}{\sqrt{n(z)}} \cdot \mathrm{e}^{\mathrm{j}\left[\omega t \mp k_0 \int^z n(z)\mathrm{d}z\right]}, \\
H_{x1,2} &= \mp \frac{A_{01,2}(z)}{Z_0} \sqrt{n(z)} \cdot \mathrm{e}^{\mathrm{j}\left[\omega t \mp k_0 \int^z n(z)\mathrm{d}z\right]},
\end{aligned} \tag{E.17}
$$

where (E.17) must be a solution of (E.18)

$$
\begin{aligned}
&E_y = E_{y1} + E_{y2}, \quad H_x = H_{x1} + H_{x2}, \\
&\frac{\partial E_{y1}}{\partial z} + \frac{\partial E_{y2}}{\partial z} = \mathrm{j}\omega\mu_0 (H_{x1} + H_{x2}), \\
&\frac{\partial H_{x1}}{\partial z} + \frac{\partial H_{x2}}{\partial z} = \mathrm{j}\omega\varepsilon_0 n^2 (E_{y1} + E_{y2}).
\end{aligned} \tag{E.18}
$$

Rearranging (E.18)

$$
\begin{aligned}
\frac{1}{A_{01}} \cdot \frac{\mathrm{d}A_{01}}{\mathrm{d}z} E_{y1} + \frac{1}{A_{02}} \cdot \frac{\mathrm{d}A_{02}}{\mathrm{d}z} E_{y2} &= \frac{1}{2n} \frac{\mathrm{d}n}{\mathrm{d}z} (E_{y1} + E_{y2}), \\
-\frac{1}{A_{01}} \cdot \frac{\mathrm{d}A_{01}}{\mathrm{d}z} E_{y1} + \frac{1}{A_{02}} \cdot \frac{\mathrm{d}A_{02}}{\mathrm{d}z} E_{y2} &= \frac{1}{2n} \frac{\mathrm{d}n}{\mathrm{d}z} (E_{y1} - E_{y2}),
\end{aligned} \tag{E.19}
$$

can be obtained. $(\mathrm{d}A_{01}/\mathrm{d}z)$ and $(\mathrm{d}A_{02}/\mathrm{d}z)$ are separable, so (E.19) can be reformulated

$$
\begin{aligned}
\frac{1}{A_{01}} \frac{\mathrm{d}A_{01}}{\mathrm{d}z} &= \frac{1}{2n} \frac{\mathrm{d}n}{\mathrm{d}z} \cdot \frac{E_{y2}}{E_{y1}}, \\
\frac{1}{A_{02}} \frac{\mathrm{d}A_{02}}{\mathrm{d}z} &= \frac{1}{2n} \frac{\mathrm{d}n}{\mathrm{d}z} \cdot \frac{E_{y1}}{E_{y2}}.
\end{aligned} \tag{E.20}
$$

As can be seen from (E.17),

$$
\frac{E_{y2}}{E_{y1}} = \frac{A_{02}(z)}{A_{01}(z)} \cdot \mathrm{e}^{\mathrm{j}2k_0 \int^z n(z)\mathrm{d}z}, \tag{E.21}
$$

and the demanded amplitude-functions are

$$\begin{aligned}\frac{dA_{01}}{dz} &= \frac{1}{2n}\cdot\frac{dn}{dz}A_{02}\cdot e^{j2k_0\int^z n(z)\cdot dz}, \\ \frac{dA_{02}}{dz} &= \frac{1}{2n}\cdot\frac{dn}{dz}A_{01}\cdot e^{-j2k_0\int^z n(z)\cdot dz}.\end{aligned} \tag{E.22}$$

(E.22) can be expressed as

$$\frac{dA_{0i}}{dz} = -\frac{1}{2n}\cdot\frac{dn}{dz}A_{0l}\cdot e^{\pm j2k_0\int^z n(z)\cdot dz}, \quad i=1,2\,;\ l=2,1\,; \tag{E.23}$$

where a solution of (E.23) is the following

$$A_{0i} = C_{0i} + \int_{z_0}^{z}\frac{1}{2n}\cdot\frac{dn}{dz}A_{0l}(z)\cdot e^{(-1)^l\cdot j2k_0\int^z n(z)\cdot dz}dz\,, \tag{E.24}$$

where z_0 is the starting point of the inhomogeneity, from which $dn/dz \neq 0$. C_{0i} = constant, that is an excitation (boundary-) condition at the place $z = z_0$.

It obviously results from the form of the solutions that the two waves transfer energy to each other, proportionally to their own amplitudes (power-density) and the intensity of the inhomogeneity (dn/dz).

Energy is reflected from both signals, the wave is attenuated on the one hand, but on the other hand, that is amplified by the absorption of the energy coupled out (reflected) from the other signal. The details depend on the exciting boundary conditions (C_{0i}).

In many cases, (E.24) in a first approximation is

$$A_{0i} \cong C_{0i} + \int_{z_0}^{z}\frac{1}{2n}\cdot\frac{dn}{dz}C_{0l}\cdot e^{(-1)^l\cdot j2k_0\int^z n\cdot dz}dz. \tag{E.25}$$

This delivers the starting value of the successive approximation. Meanwhile

$$\frac{d^2A_{0i}}{dz^2} - \left(\frac{\frac{d^2n}{dz^2}}{\frac{dn}{dz}} - \frac{1}{n}\frac{dn}{dz} \mp j2k_{0n}\right)\frac{dA_{0i}}{dz} - \frac{1}{4n^2}\left(\frac{dn}{dz}\right)^2 A_{0i} = 0 \tag{E.26}$$

is given by rearranging (E.22), so the successive approximation yields the solution of this equation.

Validity limits

The W.K.B. solution – and similarly the coupled wave-solution – is considered to be acceptably exact, while their substitution fulfils the starting equation sufficiently.

So, writing (E.16) into (E.11) it results in

$$\left[\frac{3}{4n^2}\left(\frac{\mathrm{d}n}{\mathrm{d}z}\right)^2 - \frac{1}{2n}\cdot\frac{\mathrm{d}^2 n}{\mathrm{d}z^2} - k_0^2 n^2 + k_0^2 n^2\right] E_y \triangleq 0, \tag{E.27}$$

conditional equation where, $E_y \neq 0$. The condition of acceptance of the solution is

$$\frac{3}{4n^2}\left(\frac{\mathrm{d}n}{\mathrm{d}z}\right)^2 - \frac{1}{2n}\cdot\frac{\mathrm{d}^2 n}{\mathrm{d}z^2} \cong 0. \tag{E.28}$$

That means the following relation

$$\frac{1}{k_0}\left|\frac{3}{4}\left(\frac{1}{n^2}\frac{\mathrm{d}n}{\mathrm{d}z}\right)^2 - \frac{1}{2n^3}\cdot\frac{\mathrm{d}^2 n}{\mathrm{d}z^2}\right| \ll 1. \tag{E.29}$$

(E.29) can be considered to be a quantitative definition of the weak inhomogeneity or the quasi-homogeneity.

References

Allis W.P., Buchsbaum S.J. and Bers A. (1963): Waves in Anisotropic Plasma; M.I.T. Press, Cambridge, Mass.

Anderson D., Askne J. and Lisak M. (1975): Wave packets in an absorptive and strongly dispersive medium; Phys. Rev. A., **12**, 1546.

Anderson B.J. and Fuselier S.A., (1994): Response of thermal ions to electromagnetic ion cyclotron waves; J. Geophys., Res., **99**, A10, 19.413.

Barrington R.E., Belrose J.S. and Mather W.E. (1966): A helium whistler observed in the Canadian Satellite Alouette II; Nature, **210**, 80.

Bashilov K.P., Mihajlova G.A. and Kapustina O.V. (1977): Spectrum time analysis of VLF wave-pockets by numerical methods; (in Russians), Volnovije Processi v Prizemnoj Plazme, Academia Nauk S.S.S.R., 22.

Bèghin C. and Siredey C. (1964): Un procèdè d'analyse fine des sifflements atmosphèriques; Ann. Gèophys., **20**, 301.

Bernard L.P. (1973): A new nose extension method for whistlers; J. Atmos. Terr. Phys., **35**, 871.

Berthelier J.J., Berthelier A., Galperin Yu.I., Gladyshev V.A., Lefeuvre F., Massevitch N.I., Mogilevsky M. and Molchanov O.A. (1983): Field and wave measurements aboard the AUREOL-3 spacecraft; Adv. Space Res. **2**, 49.

Bilitza D. (1990): International Reference Ionosphere 1990; National Space Science Data Center, NSSDC/WDC-A-RRS 90-20, Greenbelt, Maryland.

Bognár P., Ferencz Cs., Hamar D. and Tarcsai Gy. (1995): Whistler-mode propagation: The first results of model-calculations for a homogenous plasma; Annales Univ. Sci. Bp. R. Eötvös, Sec. Geophys. Meteor., Tom. **XI.**, 53.

Brice N.M. and Smith R.L. (1971): Whistlers: Diagnostic tools in space plasma; in Methods of Experimental Physics, **9B**, Academic Press, New York.

Brillouin L. (1960): Wave Propagation and Group Velocity; Academic Press, San Diego, Calif. (First publ. in 1914.)

Budden K.G. (1966): Radio waves in the ionosphere; Cambridge Univ. Press, London.

Budden K.G. (1968): Lectures on Magnetoionic Theory; In Geophysics, The Earth's Environment, Gordon and Breach, New York, 90.

Bullough K. and Sagredo J.L. (1973): VLF Goniometer Observations at Halley Bay, Antarctica - I. The Equipment and the Measurement of Signal Bearing; Planet Space Sci. **21**, 899.

Cain J.C., Shapiro I.R. and Stolarick J.D. (1962): Whistler signals observed with the Vanguard III satellite; J. Phys. Soc. Japan, **17**, 84.

Carpenter D.L. and Seely N.T. (1976): Cross-L plasma drifts in the outer plasmasphere: quiet time patterns and some substorm effects; J. Geophys. Res., **81**, 2728.

Chappell C.R., Harris K.K. and Sharp G.W. (1970): The morphology of the bulge region of the plasmasphere; J. Geophys. Res., **75**, 3848.

Cook C.E. (1960): Pulse compression-key to more efficient radar transmission; Proc. I.R.E. (USA), **48**, 310.

Coroniti F.V., Fredericks R.W., Kennel C.F. and Scarf F.L. (1971): Fast time resolved spectral analysis of VLF banded emissions; J. Geophys. Res., **76**, 2366.

Cöster C. (1974): Calibration of a VLF goniometer receiver and development of a method for computer processing of whistlers; ESA-ESTEC, Nordwijk.

Cserepes L. (1987): Notes on synthesizing ion-cyclotron whistlers by the full-wave method; Ann. Geophys., **5A** (3), 155.
Daniell G.J. (1969): The representation of signals in frequency and time; (Manuscript), 1-13.
Daniell G.J. (1970): The representation of signals in frequency and time; (Manuscript), 1-23.
Daniell G.J. and Papanicolau C. (1975): Calculations with digital sonograms; URSI XVIIIth Gen. Ass., Lima, Abstracts, 21.
DiFranco J.V. and Rubin W.L. (1968): Radar Detection; Prentice-Hall, Englewood Cliffs.
Dziewonsky A., Bloch S. and Landisman M. (1969): A technique for the analysis of transient seismic signals; Bull. Seism. Soc. Am., **59**, 427.
Elie F., Hayakawa M., Parrot M., Pincon J.L. and Lefeuvre F. (1999): Neural network system for the analysis of phenomena on board the DEMETER micro-satellite; IEICE Trans. Fundamentals, **E82-4**, 1578.
Ferencz Cs. (1977a): Electromagnetic wave propagation in inhomogeneous media: Strong and weak inhomogeneities; Acta Technica Ac. Sci. H., **85(3-4)**, 433.
Ferencz Cs. (1977b): A geometric resolution of the contradiction between the propagation of electromagnetic plane wave in moving dielectrics and the Einsteinian addition of velocities; Acta Technica Ac. Sci. H., **84(1-2)**, 147.
Ferencz Cs. (1978a): Electromagnetic wave propagation in inhomogeneous media: Method of inhomogeneous basic modes; Acta Technica Ac. Sci. H., **86(1-2)**, 79.
Ferencz Cs. (1978b): Electromagnetic wave propagation in inhomogeneous media: The analysis of the group velocity; Acta Technica Ac. Sci. H., **86(1-2)**, 169.
Ferencz Cs. (1987): Geometric questions of electromagnetic wave propagation; Acta Technica Ac. Sci. H., **100(3-4)**, 195.
Ferencz Cs. (1995a): Whistler-mode propagation: Solution for a homogeneous plasma; Annales Univ. Sci. Bp. R. Eötvös, Sec. Geophys. Meteor., Tom. **XI.**, 5.
Ferencz Cs. (1995b): Whistler-mode propagation: Solution for an inhomogeneous plasma; Annales Univ. Sci. Bp. R. Eötvös, Sec. Geophys. Meteor., Tom. **XI.**, 27.
Ferencz Cs. (1996): Electromagnetic wave propagation; (in Hungarian) Academic Press, Budapest, Hungary.
Ferencz Cs., Bognár P., Tarcsai Gy., Hamar D. and Smith A.J. (1996): Whistler-mode propagation: results of model calculations for an inhomogeneous plasma; J. Atmos. Terr. Phys., **58**, 625.
Ferencz O. (1993): The Whistler-mode Propagation: Solution for a Lossy Magnetoionic Medium. URSI XXIVth Gen. Ass., Kyoto, H6.p-04.
Fijalkow E., Altman C. and Cory H. (1973): A full-wave study of ion-cyclotron whistlers in the ionosphere; J. Atmos. Terr. Phys., **35**, 317.
Gábor D. (1946): Theory of communication; J. IEE (London), **93**, 429.
Gendrin R. and de Villedary C. (1979): Apparent and real fine structures in magnetospheric ULF or VLF emissions; J. Geophys. Res., **84**, 5839.
Griffiths L.J. (1975): Rapid measurements of digital instantaneous frequency; IEEE Trans. Acoust., Speech Signal Process., **ASSC-23**, 207.
Gonzales C.A., Kelley M.C., Carpenter D.L., Miller T.R., and Wand R.H. (1980): Simultaneous measurements of ionospheric and magnetospheric electric fields in the outer plasmasphere; Geophys. Res. Lett., **7**, 6127.
Gurnett D.A., Shawhan S.D., Brice N.M. and Smith R.L. (1965): Ion Cyclotron Whistlers; J. Geophys. Res., **70**, 7/1665.
Gurnett D.A and Shawhan S.D. (1966): Determination of Hydrogen Ion Concentration, Electron Density, and Proton Gyrofrequency from the Dispersion of Proton Whistlers; J. Geophys. Res., **71** , 3/741

Gurnett D.A., Kurth W.S., Cairns I.H. and Granroth L.J. (1990): Whistlers in Neptune's Magnetosphere: Evidence of Atmospheric Lightning; J. Geophys. Research, **95**, A12, 20.967.

Hamar D. and Tarcsai Gy. (1982): High resolution frequency-time analysis of whistlers using digital matched filtering. Part I.: Theory and simulation studies; Annales Geophys. **38**, 119.

Hamar D., Tarcsai Gy., Lichtenberger J., Smith A.J. and Yearby K.H. (1990). Fine structure of whistlers recorded digitally at Halley, Antarctica; J. Atm. Terr. Phys., **52**, 801.

Hamar D. and Tarcsai Gy. (1991): Matched filtering of whistlers; Ann. Univ. R. Eötvös Geoph. Meteor., Tom. **VI-VII**, 51.

Hamar D., Ferencz Cs., Lichtenberger J., Tarcsai Gy., Smith A.J. and Yearby K.H. (1992): Trace splitting of whistlers: A signature of fine structure or mode splitting in magnetospheric ducts?; Radio Science, **27**, 341.

Havelock T.H. (1914): The Propagation of Waves In Dispersive Media; Cambridge University Press, New York.

Hayakawa M. and Fujinawa Y. (1994): Electromagnetic Phenomena Related to Earthquake Prediction, Ch.4. Satellite Observation of Seismogenic Phenomena and Perspective Methods; Terra Scientific Publ. Co. Tokyo.

Hayakawa M. (1999): Atmospheric and Ionospheric Electromagnetic Phenomena Associated with Earthquakes; Terra Scientific Publ. Co. Tokyo.

Heald M.A. and Wahrton C.B. (1965): Plasma diagnostics with microwaves; John Wiley & Sons, New York.

Helliwell R.A. and Gehrels E. (1958): Observation of magneto-ionic duct propagation using man-made signals of very low frequency; Proc. IRE, **46**, 785.

Helliwell R.A. (1965): Whistlers and Related Ionospheric Phenomena; Stanford University Press, Stanford.

Helstron C.W. (1968): Statistical Theory of Signal Detection; Pergamon Press, Oxford.

Holden D.N., Munson C.P. and Devenport J.C. (1995): Satellite observations of transionospheric pulse pairs; Geophys. Res. Lett., **22**, 889.

Horne R.B. and Thorne R.M. (1994): Convective instabilities of electromagnetic ion cyclotron waves on the outer magnetosphere; J. Geophys. Res., **99**, A9, 17.259.

Hudson D.J. (1970): Statistics; (Statistika dlia fizikov; Russian translation 2nd edition, MIR Publishers, Moscow.)

Idemen M. (1973): The Maxwell's Equations in the sense of distributions; IEEE Trans. on Ant. and Prop., **AP-18**, 242.

Jiricek F. and Tarcsai Gy. (1980): Analysis of whistlers observed simultaneously aboard the Intercosmos-10 satellite and on the ground; Acta Geod. Geophys. Montan. Acad. Sci. Hung., **15**, 97.

Jones D. (1969): The effect of the latitudinal variation of the terrestrial magnetic field strength on ion cyclotron whistlers; J. Atm. Terr. Phys., **31**, 971.

Kaiser M.L., Desch M.D., Warwick J.W. and Pearce J.B. (1980): Voyager detection of nonthermal radio emission from Saturn; Science, **209**, 1238.

Kamke E. (1956): Differentialgleichungen I.; Geest & Portig, Leipzig (Punkt 2.35, pp. 412-413.).

Kamke E. (1965): Differentialgleichungen II.; Geest & Portig K.G., Leipzig (Punkt 11.1, p. 93.).

Karpman V.I. and Kaufman R.N. (1981): Tunneling transformation of whistler waves in an inhomogeneous plasma; Sov. Phys. JETP, **53** (5), 956.

Karpman V.I. and Kaufman R.N. (1982a): Whistler wave propagation in the density ducts; J. Plasma Phys., **27** (2), 225.
Karpman V.I. and Kaufman R.N. (1982b): General Schrödinger equation for whistler waves propagating along a magnetic field; Phys. Lett., **90**, 137.
Karpman V.I. and Kaufman R.N. (1987): Whistler wave propagation in plasma ducts; Radio Sci., **22**, 1023.
Karpman V.I. and Shagalov A.G. (1989): Influence of dispersion on the modulational instability of a whistler wave; J. Plasma Physics, **41**, 289.
Kimura I., Matsumoto H., Mukai T., Hashimoto K., Helliwell R.A., Bell T.F., Inan U.S. and Katsufrakis J.P. (1980): JIKIKEN (EXOS-B) observation of Simple transmissions; COSPAR Symp. on Active Exp. in Space Plasmas, Budapest.
Klauder J.R., Price A.S., Darlington S. and Albersheim W.J. (1960): The theory and design of chirp radars; The Bell System Techn. Journ., **39**, 745.
Laird M.J. and Nunn D. (1975): Full wave VLF modes in a cylindrically symmetric enhancement of plasma density; Planet. Space Sci., **23**, 1649.
Laird M.J. (1992): Mode theory of whistler ducts: integrated group delay times; J. Atmos. Terr. Phys., **54**, 1599.
Laue V.M. (1955): Die Relativitätstheorie, I.; Vieveg und Sohn, Braunschweig.
Leavitt M.K. (1975): A frequency-tracking direction-finder for whistlers and other very-low-frequency signals; Ph.D. thesis, Radioscience Lab., Stanford University, Stanford, California.
Leer E., Johansen K.M. and Albrigtsen R. (1978): Group velocity of whistlers in a two-ion plasma; J. Geophys. Res., **83**, A7, 3127.
Lerner R.M. (1961): Representation of signals; Communication System Theory, New York, McGraw-Hill, 203.
Levshin A.L., Pisarenko V.F. and Pogrebinsky G.A. (1972): On a frequency-time analysis of oscillation; Ann. Geophys., **28**, 211.
Lichtenberger J., Hamar D. and Cserepes L. (1987): Computation of whistler wave normals using a combined matched filtering and parameter estimation technique; J. Atmos. Terr. Phys., **49**, 1075.
Lichtenberger J., Tarcsai Gy., Pásztor Sz., Ferencz Cs., Hamar D., Molchanov O.A. and Golyavin A.M. (1991): Whistler doublets and hyperfine structure recorded digitally by the Signal Analyzer and Sampler on the Active satellite; J. Geophysical Research, **96**, A12, 21.149.
Maeda K. and Kimura I. (1956): A theoretical investigation of the propagation path of the whistling atmospherics; Rept. Iono. Space Res. Japan, **10**, 105.
Massey R.S. and Holden D.N. (1995): Phenomenology of transionospheric pulse pairs; Radio Sci., **30**, 1645.
Means J.D. (1972): Use of three-dimensional covariance matrix in analyzing the polarization properties of plane waves; J. Geophys. Res., **77**, 5551.
Menietti J.D., Tsintikidis D., Gurnett D.A. and Curren D.B. (1991): Modelling of Whistler Ray Paths in the Magnetosphere of Neptune; J. Geophys. Research, **96**, 19.117.
Mihajlova G.A. (ed.) (1977): Volnovie processi i prizemnoi plasme; Acad. Nauk USSR, Moscow (pp. 22-49).
Mihajlova G.A. and Kapustina O.V. (1978): Spectrum-time analysis of whistlers by numerical methods; Geomagnet. i Aeronom., **18**, 473, (in Russian).
Mihovilović D. and Bracewell R.N. (1992): Whistler analysis of the time-frequency plane using chirplets; J. Geophys. Res., **97**, A11, 17.199.

Miniere X., Pircon J.L. and Lefeuvre F. (1995): A neural network approach to the classification of electron and proton whistlers; J. Atmos Terr. Phys. **58**, 911.
Morgan M.G. and Allock G.M. (1956): Observations of whistling atmospherics at geomagnetically conjugate points; Nature, **177**, 30.
Okada T., Iwai A. and Hayakawa M. (1977): The measurement of incident and azimuthal angles and the polarization of whistlers at low latitudes; Planet Space Sci., **25**, 233.
Osterberg H. (1958): Propagation of plane electromagnetic waves in inhomogeneous media; J. Opt. Soc. Ann., **48**, 513.
Park C.G. (1972): Methods of determining electron concentrations in the magnetosphere from nose whistlers; Technical Report 3454. Radiosci. Lab., Stanford Univ., Stanford, (pp. 11-17 and p. 23).
Parrot M. (1994): Statistical study of ELF/VLF emissions recorded by a low-altitude satellite during seismic events; J. Geophys. Research, **99**, A12, 23.339.
Rihaczek A.W. (1969): Principles of High-resolution Radar; McGraw Hill, New York.
Rishbeth H. and Garriott O.K. (1969): Introduction to ionospheric physics; Int. Geophys. Series. Vol. 14., Academic Press, New York.
Rodger C.J. (1999): Red sprites, upward lightening, and VLF perturbations; Reviews Geophys., **37**, 317.
Rodriguez P. and Gurnett D.A. (1971): An Experimental Study of Very-Low-Frequency Mode Coupling and Polarization Reversal; J. Geophys. Res., **76**, 4/960.
S-300 Experimenters (1979): Measurements of electric and magnetic wave fields and of cold plasma parameters on board GEOS-1, Preliminary results; Planet. Space Sci., **27**, 317.
Sagredo J.L. and Bullough K. (1973): VLF goniometer observations at Halley Bay, Antarctica - II. Magnetospheric structure deduced from whistler observations; Planet Space Sci., **21**, 913.
Sazhin S.S., Hayakawa M. and Bullough K. (1992a): Whistler diagnostics of magnetospheric parameters: a review; Ann. Geophys., **10**, 293.
Sazhin S., Smith A.J., Bullough K., Clilverd M.A., Saxton J.M., Strangeways H.J. and Tarcsai Gy. (1992b): Group delay times of whistler-mode signals from VLF transmitters observed at Faraday, Antarctica; J. Atm. Terr. Phys., **54**, 99.
Sazhin S. (1993): Whistler-mode Waves in a Hot Plasma; Cambridge Univ. Press, Cambridge.
Scarabucci R.R. and Smith R.L. (1971): Study of magnetospheric field oriented irregularities the mode theory of bell-shaped ducts; Radio Sci., **6**, 65.
Shawhan S.D. (1983): Estimation wave vector characteristic; Adv. Space Res., **7**, 31.
Simonyi K. (1963): Foundations of Electrical Engineering; Pergamon Press, New York.
Singh Ashok K., Singh Abhay K., Singh D.K. and Singh R.P. (1998): The effect of temperature on the dispersion of proton whistlers; J. Atm. Sol.-Terr. Phys., **60**, 551.
Siredey C. (1967): Un système de dèpouillement numèrique des sifflements atmosphèriques; D. Ing. Thesis, Universitè de Paris.
Smith A.J. (1995): VELOX: a new VLF/ELF receiver in Antarctica for the Global Geospace Science mission; J. Atm. Terr. Phys., **57**, 507.
Smith E.J., Davis L., Jones D.E., Coleman P.J., Colburn D.S., Dyal P. and Sonett C.P. (1980): Saturn's magnetic field and magnetosphere; Science, **207**, 407.
Smith R.L. (1960): Guiding of whistlers in a homogeneous medium; J. Res. NBS-D, Radio Propagation, **64D** 505.
Smith R.L., Helliwell R.A. and Yabroff I. (1960): A theory of trapping of whistlers in field-aligned columns of enhanced ionization; J. Geophys. Res., **65**, 815.

Smith R.L. and Brice N. (1964): Propagation in Multicomponent Plasmas; J. Geophys. Res. **69**, 23/5029.
Sonwalkar V.S. and Inan U.S. (1986): Measurements of Simple transmitter on DE-1 satellite: wave normal direction and antenna effective length; J. Geophys. Res., **91**, 154.
Sonwalkar V.S. and Inan U.S. (1988): Wave normal direction and spectral properties of whistler-mode hiss observed on the DE-1 satellite; J. Geophys. Res., **93**, 7493.
Sonwalkar V.S., Carpenter D.L. and Strangeways H.J. (1991): Testing Radio Burst Observed on the Nightside of Venus for Evidence of Whistler Mode Propagation from Lightning; J. Geophys. Research, **96**, A10, 17.763.
Stiles G.S. and Helliwell R.A. (1975): Frequency-time behavior of artificially stimulated VLF emissions; J. Geophys. Res. **80**, 608.
Stix T.H. (1962): Theory of plasma waves; McGraw-Hill, New York.
Storey L.R.O. (1953): An investigation of whistling atmospherics; Phil. Trans. R. Soc., Ser. **A246**, 113.
Storey L.R.O. (1957): A method for interpreting the dispersion curves of whistlers; Can. J. Phys., **35**, 1107.
Storey L.R.O. and Lefeuvre F. (1979): The analysis of 6-component measurements of a random electromagnetic wave field in a magnetoplasma - I. The direct problem; Geophys. J. R. Astr. Soc., **56**, 255.
Storey L.R.O. and Lefeuvre F. (1980): The analysis of 6-component measurements of a random electromagnetic wave field in a magnetoplasma - II. The integration kernel; Geophys. J. R. Astr. Soc., **62,** 173.
Storey L.R.O., Lefeuvre F., Parrot M., Cairo L. and Anderson R.R. (1991): Initial survey of the Wave Distribution Functions for plasmaspheric hiss observed by ISEE1; J. Geophys. Res., **96,** 19.469.
Strangeways H.J. (1977): Ph.D. Thesis; Univ. of Southampton.
Strangeways H.J. (1980): Systematic errors in VLF direction-finding of whistler ducts, I; J. Atmos. Terr. Phys., **42**, 995.
Strangeways H.J. and Rycroft M.J. (1980): Systematic errors in VLF direction-finding of whistler ducts, II; J. Atmos. Terr. Phys., **42**, 1009.
Strangeways H.J. (1982): The effect of multiduct structure on whistler-mode wave propagation; J. Atmos. Terr. Phys., **44**, 901.
Suchy K. (1972): The velocity of a wave packet in an anisotropic absorbing medium; J. Plasma Physics, **8**, 33.
Tanaka M., Fujiwara M. and Ikegami H. (1986): Propagation of a Gaussian wave packet in an absorbing medium; Phys. Rev. A., **34**, 4851.
Tarcsai Gy. (1975): Routine whistler analysis by means of accurate curve fitting; J. Atmos. Terr. Phys., **37**, 1447.
Tarcsai Gy. (1981): Two new procedures for whistler analyzing; (in Hungarian) CSc thesis on Hung. Acad. Sci., Budapest.
Tarcsai Gy., Strangeways H.J. and Rycroft M.J. (1989): Error sources and travel time residuals in plasmaspheric whistler interpretation; J. Atmos. Terr. Phys., **51**, 249.
Tarcsai Gy., Molchanov O., Hamar D., Lichtenberger J., Shapiro V. and Shevchenko V. (1992): Observation of narrow-band changing tone emissions during operation of VLF emitters onboard Activny satellite; The World Space Congress, Washington D.C., D.5-M.1.12X, 447.
Terry P.D. and Budden K.G. (1978): A complex ray tracing study of ion cyclotron whistlers in the ionosphere; Proc. R. Soc. Lond. A., **363**, 425.

Thomson R.J. and Dowden R.L. (1977a): Simultaneous ground and satellite reception of whistlers, I. Ducted whistlers; J. Atm. Terr. Phys., **39**, 869.
Thomson R.J. and Dowden R.L. (1977b): Simultaneous ground and satellite reception of whistlers, II. Non-ducted whistlers; J. Atm. Terr. Phys., **39**, 879.
Thorne R.M. and Horne R.B. (1994a): Landau damping of magnetospherically reflected whistlers; J. Geophys. Res., **99**, A9, 17.249.
Thorne R.M. and Horne R.B. (1994b): Energy transfer between energetic ring current H^+ and O^+ by electromagnetic ion cyclotron waves; J. Geophys. Res., **99**, A9, 17.275.
Tixier M. (1976): Contribution á l`ètude de la propagation des ondes radioèlectriques de très basse frèquence rayonnèes par l`èclair; Ph.D. thesis, Universitè de Poitiers.
Tsuruda K. and Hayashi K. (1975): Direction finding technique for elliptically polarized VLF electromagnetic waves and its applications to the low-latitude whistlers; J. Atmos. Terr. Phys., **37**, 1193.
Vilalón E. and Bruke W. (1993): Proton Whistler Interactions Near the Equator in the Radiation Belts; J. Geophys. Res., **98**, A8, 13.509.
Walker A.D.M. (1976): The theory of whistler propagation; Revs. Geophys. Space Phys., **14**, 629.
Walker A.D.M. (1993): Plasma waves in the magnetosphere; Phys. Chem. Space Planet., **24**, Springer-Verlag, Berlin.
Wang Ting-I. (1971): Intermode Coupling at Ion Whistler Frequencies in a Stratified Collisionless Ionosphere; J. Geophys. Res., **76**, 947.
Watanabe S. and Ondoh P. (1976): Deuteron-whistler and trans-equatorial propagation of the ion-cyclotron whistler; Planet. Space Sci., **24**, 359.
Woodward P.M. and Davies I.L. (1950): A theory of radar information; Phil. Mag., **41**, 1001.
Wong H.K. and Goldstein M.L. (1987): Proton Beam Generation of Whistler Waves in the Earth's Foreshock; Geophys. Res., **92**, A11, 12.419.
Xu J.S. and Yeh K.C. (1990): Propagation of a VLF electromagnetic wave packet in a magnetoplasma; J. Geophys. Res., **95**, A7, 10.481.
Yarboff L. (1961): Computation of whistler ray paths; J. Res. NBS., **65D**, 485.
Zadeh L.A. and Ragazzini J.R. (1952): Optimum filters for the detection of signals in noise; Proc. IRE, **40**, 1123.

Subject Index

Made in the USA
Las Vegas, NV
13 November 2024

11571135R00151